# THE POLITICS
# OF POLLUTION
# IN A COMPARATIVE
# PERSPECTIVE

Carl Beck, Frederic J. Fleron, Jr.,
Milton Lodge, Derek J. Waller,
William A. Welsh, M. George
Zaninovich, Comparative Communist
Political Leadership

Roger W. Benjamin (with Alan Arian,
Richard N. Blue, and Stephen
Coleman), Patterns of Political
Development: Japan, India, Israel

Cynthia H. Enloe, The Politics of
Pollution in a Comparative
Perspective: Ecology and Power in
Four Nations

Wolfram F. Hanrieder (ed.),
Comparative Foreign Policy:
Theoretical Essays

Martin O. Heisler (ed.), Politics in
Europe: Structures and Processes in
Some Postindustrial Democracies

Gregory Henderson, Richard Ned
Lebow, and John G. Stoessinger (eds.),
Divided Nations in a Divided World

Allan Kornberg (ed.), Legislatures in
Comparative Perspective

Stein Rokkan (with Angus Campbell,
Per Torsvik, and Henry Valen),
Citizens, Elections, Parties:
Approaches to the Comparative Study
of the Processes of Development

Jonathan Wilkenfeld (ed.), Conflict
Behavior and Linkage Politics

*Comparative Studies
of Political Life*
SERIES EDITOR:
**Martin O. Heisler**

# THE POLITICS OF POLLUTION IN A COMPARATIVE PERSPECTIVE:

Cynthia H. Enloe
*Clark University*

## Ecology and Power in Four Nations

DAVID McKAY COMPANY, INC.
New York

**To David,**

**who thrives on fresh air
and cities**

The Politics of Pollution in a
Comparative Perspective: Ecology and
Power in Four Nations

COPYRIGHT © 1975 BY DAVID MCKAY
COMPANY, INC.

International Standard Book Number:
0–679–30275–1 (paper); 0–679–30279–4
(cloth)

LIBRARY OF CONGRESS CATALOG CARD
NUMBER: 75–888

MANUFACTURED IN THE UNITED STATES
OF AMERICA

*Design by Bob Antler*

# ACKNOWLEDGMENTS

No study of this scope can be undertaken without the assistance of generous colleagues. One's dependence on others is even greater when the topic is in as much flux as is environmental policy. Among those who have patiently tried to introduce me to the scientific intricacies of environmental control, I am particularly indebted to my colleagues at Clark University, Roger Kasperson, Robert Kates, Len Berry, and Michael McClintock.

I was given helpful suggestions and criticisms by Donald Kelley, James Dunn, T. J. Pempel, Norman Vig, Julian Cummins, David Withrington, Ian Burton, and Les Foster.

The Harrington Research Fund of Clark University provided financial support during part of this project.

For their skill and patience I am also indebted to Terry Reynolds and her typing staff at Clark University.

Responsibility for interpretations as well as errors or omissions, must, of course, be mine.

# FOREWORD

When this series was inaugurated in 1970, its purpose was given in the following terms: "to combine the theoretical and empirical fruits of the best and most modern scholarship in the field of comparative politics and present them in readily intelligible and usable form." Two unspoken hopes accompanied the expressed aim: (1) that the vocational noises and necessary but distracting scaffolding of such scholarship be removed from hearing and sight—to avoid what for most readers is, after all, obtrusion; and, more important, (2) that the subjects to which the scholarship was applied be truly significant problems in "the real world." These silent goals seemed distant indeed; for, while the series was launched at a time when the field was emerging from two decades of vital but distracting internal restructuring, the application of the gains that came from that long period of inelegant and sometimes involuted development to important problems in the political lives of societies appeared somewhat remote. It is a measure of the

vi

collective achievements of social science and more particularly in this instance of Cynthia Enloe, rather than an
affected gesture of self-congratulation, that such application is now possible; and it can be said to have been begun
in earnest with the publication of this volume.

As with most other good books, several levels of
meaning can be discerned in this work. For most readers,
the book is what it claims to be: a comparative study of
politics and public policy surrounding one of the crucial
questions of the present and the future—environmental
pollution. In addition to the in-depth studies of four
important countries, Professor Enloe presents accurate
and informative sketches of environment-related problems and policy processes in numerous other settings.

The book is quite comprehensive in analytic as well
as geographic scope: all important facets of the political
processes surrounding environmental pollution are
treated, from the crucial stage of "politicizing" the problem through mobilization, policy making, and implementation to a concern with policy impacts. Such breadth of
concern is vital if the comparative aspects of the work are
to retain meaning and accuracy *in context;* for it should
be evident that whether or how or how quickly a fact of
life becomes a political issue to be treated through the
machinery of public policy is no less important for
determining the outcome than the operation of that
machinery. At this broadest level, what follows is a
comprehensive comparative study of the politics of pollution.

There has been a great deal of discussion in the social
sciences during the past decade about "public policy"—
what it is and how it should be studied, and how truly
comparative analyses might be undertaken. The preliminary discussions about concepts, methods, and pretheo-
ries have been useful as well as necessary. One hardly
needs to observe that they are insufficient. Sooner or later
those who are interested in the making and implementa-

tion of the authoritative decisions—that affect (1) distributions and uses of large portions of our resources and (2) our values—had to turn to substantive studies. This book is an excellent venture into what is still a largely uncharted realm. Professor Enloe's concerns are varied and comprehensive from the perspectives of analysis and theory building. Thus, she treats the manner in which environmental questions move from the arena of private to public decisions, as well as their disposition in the public sector. Environmental policy is shown to be sometimes "regulative" and sometimes "self-regulative," to use the jargon of public policy analysis. That is, public policy regarding environmental matters takes the form of either bureaucratized regulation or injunctions and mandates for self-regulation by the private sector. This distinction tells us much about the political systems concerned. Pollution problems that are not subjected to some form of public treatment often result in "redistribution" —another category of policy outputs—*from the less wealthy to the more wealthy,* since the former often bear the costs and burdens that accompany pollution while the latter reap higher profits.

This, then, is an important book both because it concerns one of the modern world's most crucial problems as it is manifested and dealt with in some of the most important countries and because it provides us with an applied and broadly comparative study of public policy from issue inception through policy making and implementation to policy impact. For me it has another, considerable, virtue: it demonstrates to any who might have doubted it that modern social scientific scholarship *can* be applied to our most pressing problems—and that it can be done with grace and in readily intelligible terms.

*Martin O. Heisler*

# CONTENTS

# INTRODUCTION

Almost by its very nature politics involves tension, though the object of politics is resolution. The irony of the present era, however, is that politics is preoccupied with tensions derived in large part from its own success. The capacity of political institutions to mobilize men and resources in nation-states for the sake of increasing gross economic output is one of the success stories of political history. What makes the political tensions generated by environmental hazards so discomforting is not their violence (environmental debates have produced no wars or riots) but, rather, their implicit questioning of the very political formulas that have bestowed legitimacy on institutions and ideologies in a wide variety of systems.

A Japanese fisherman's daughter, age ten, caught this

spirit of unease in the midst of politically designed progress:

> When my father pulled in the net,
> faded seaweeds hung down from his hands,
> His lips moved slowly when he said,
> "This is the last time I'll pull the net this year."
> The highway runs across the pools
> where the seaweeds grow.
> Everybody likes the highway
> but it's made a crack
> between the ocean
> and my father
> and my mind.[1]

Most studies that have explored the dynamics of environmental politics have concentrated on the United States. But, as the Japanese girl's verse suggests, other political systems are having to cope with environmental issues as well. It would be a mistake to assume that environmental hazards have the same issue saliency in other countries as they do in the United States—in some, the issue enjoys greater primacy; in others, it is scarcely on the political stage. Furthermore, it would be risky to assume that even where they have achieved issue status environmental hazards are subjected to the same pressures, institutional fragmentations, and ideological presumptions that shape their political handling in the United States. One of the principal purposes of this book is to examine environmental debates and policies in countries outside the United States, not only for the sake of expanding our knowledge about the varieties of politi-

---

1. Suyama Sadae, in *There Are Two Lives: Poems by Children of Japan*, ed. Richard Lewis (New York: Simon & Schuster, 1970). Reprinted by permission of the author and Akane Book Company, Tokyo.

cal approaches to environmental control, but also in order
to gain perspective on the American approach, which all
too often is treated in splendid isolation.
This book is written with two sets of audiences in
mind. The first is composed of persons primarily concerned about halting environmental deterioration flowing
from the deliberate acts of men. For these persons,
perhaps the role to assume while digesting the subsequent chapters is that of a Ralph Nader or Barry Com-
moner, two of the foremost American environmental
lobbyists, on a world political tour. How much would
these American political activists have to adjust their
arguments and strategies as they moved from Poland to
Indonesia to Japan? Environmental problems do not
confine themselves to neat legal boundaries of nation-
states. Polluted air from one country blows across to
make breathing difficult for citizens in another country.
Waste dumped by corporations or cities in one nation
flows through riverways to jeopardize drinking water in a
neighboring nation. Supertankers flying one country's
flag deliberately or accidentally leak their oil cargoes into
coastal waters of the several nations they pass in their
long voyages. The globe-trotting environmentalist becomes a more and more common figure as environmental
problems require political decision making not by nations
individually, but collectively. Such a political process is
not merely a matter of international relationships. Internal, domestic political dynamics of individual nations will
be crucial and thus have to be comprehended in a
sophisticated manner by American environmentalists
working for international safeguards.
The best equipment for the American environmentalist to carry in his or her baggage when moving from
country to country is a set of questions that alert the
activist to the pertinent differences in policy formation.
No one can be expected to know the intricacies of

cultural, party, or bureaucratic politics of some 130 sovereign nations. Nevertheless, one can be expected to possess an empirical alertness and analytical systematism that reduces American political parochialism. The following chapters are intended, therefore, to suggest the sorts of questions that can elicit the most accurate and meaningful insight into a wide number and variety of political systems as they adopt or resist environmental controls.

The second audience for which this volume is written includes students of politics, whether or not they are particularly interested in environmental affairs. There are numerous "handles" that one can employ to extract analytical knowledge out of confused political reality. One can compare one entire political system with another or single institutions, such as courts or parties, as they function in several polities. More recently, comparative political analysts have preferred to focus on behavioral phenomena that seem to crop up in disparate systems: revolutions, political violence, elections, corruption, ethnic separatism. In each approach the goal is the same— not only to say something meaningful about the specific topic but also to contribute testable generalizations about the wider conditions that shape how men and women utilize power to make public, collective choices.

Another approach with the same objective is selection of a single policy area or issue—urbanization, health care, land reform, taxation—asking how it is handled in a number of different political systems. The environmental issue was chosen in the present comparative study for several reasons. First, it is a policy realm that only recently has achieved the status of a political issue in most countries and thus permits us to explore the dynamics of issue creation, rather than simply issue resolution. Second, environmental problems are subject to political "sprawl" and thus touch cultural and ideological nerves in a political system, rather than simply organized interests or institutionalized governmental processes. Third, it is an issue that has gained political prominence in communist as well as noncommunist systems and in industrialized as well as underdeveloped countries. Thus, using the environmental issue for the sake of comparison prompts us to overcome those analytical dichotomies that so often obstruct interesting comparative research. As the following pages demonstrate, there appear to be as many similarities as differences between communist and noncommunist systems in their treatment of environmental matters. There also appear to be important ways in which "underdevelopment" plagues not only the countries of Asia and Africa, but also affluent nations which, despite their lofty GNPs, are underdeveloped insofar as they lack the cultural and institutional capacity to formulate and implement effective environmental controls.

Finally, the environmental issue as a focus for comparative research has the advantage of shedding light on the workings of government bureaucracies. Protection of air, water, wildlife, and landscape all require bureaucratic implementation and monitoring. Moreover, in most countries established bureaucratic agencies have preexisting stakes in certain economic arrangements or public programs which lead them to be active participants in making as well as implementing environmental policies. Although public administration experts have been urging us to look closely at the administrative side of political systems, they have generally been ignored in comparative studies. It has only been since we have noticed the impact that intraagency rivalries have on the making of governmental policy that we have begun to devote attention to bureaucratic politics. Bureaucracies do not, of course, play the same role or carry equal weight in all systems. Yet, investigation of environmental issues makes clear that variations of bureaucratic cohesiveness and politicization are a critical factor determining whether a nation is receptive to environmental questions and whether it is effective in carrying out environmental laws. The book is divided into two parts. The first four chapters raise general analytical questions concerning how environmental hazards become political issues and how they are resolved. Illustrations are drawn from a wide assortment of political systems—Canada, China, France, Brazil, Sweden. No country is dealt with in its entirety. Yet it is hoped that these chapters will generate questions that can be utilized in any country in order to describe and explain the dynamics of environmental politics or, for that matter, governmental *neglect* of environmental problems. An underlying presumption in this first section is that environmental disruptions do not gain political attention automatically. They must occur in conjunction with certain political factors if they are to be raised to the status of a serious question of policy.

The second half of the volume focuses on four countries in particular and describes in some detail how their political systems have coped with growing man-made threats to the environment. The countries chosen for individual treatment are the United States, the Soviet Union, Japan, and Britain. They offer contrasts in cultural and ideological bases of politics, in sequence of modernization, in relationship between state and private authority over the economy, and in degree of centralized political power. Each is important internationally in offering aid and models to other countries searching for ways to resolve the problems caused by modernization. Finally, because of their international influence each attracts considerable attention from social scientists and journalists so that we have more data available on them in any policy area than we do on most other countries. All these factors make them interesting and useful objects for case-study analysis.

Despite their disproportionate economic and political power, each of the four countries singled out for special attention can be analyzed using the questions raised in the earlier comparative chapters. In this sense, the two sections of the book should flow back and forth. In the future, if environmental protection is to be truly effective internationally, we are going to need case studies of this character for a multitude of nations. This should not only assist in environmental policy making, but also contribute to our analytical sophistication generally.

The subject of the present study is *politics*. Ethics, ecological science, and economics all are critical approaches to environmental affairs; but they are included here only as they become political factors in a nation's policy process. One of the exciting qualities of environmental problems is their absorption of so many dimensions of human experience. Politics is only one such dimension. But as man-made disruptions of his physical habitat become increasingly harmful it is the political arena in which all those dimensions meet. Power—its forms, distribution, and justifications—not only has played a major role in creating environmental hazards, but also will determine how and whether those hazards are controlled.

# COMPARATIVE ANALYSIS

PART I

# HOW THE ENVIRONMENT BECOMES AN ISSUE

## CHAPTER 1

## "ISSUENESS"

Most studies of political power and policy making concentrate on the process through which public issues are resolved. The conventional focus is on how public demands gain access to centers of power, how conflicting demands are reconciled and governmental rewards and sanctions are administered. All three stages of political activity—articulation, decision, and implementation—are crucial to a nation's ability to cope with problems of environmental pollution. But prior to them is another political process: *the transformation of a fact of life into a political issue.* Environmental pollution offers an unusual opportunity to dissect this process because it has

11

only recently gained the status of an "issue" in most countries.

One reason for the neglect of the process of issue creation is that it is so difficult to pin down. It does not occur at a designated time the way an election does; it does not involve a specific set of actors; there isn't even a solid definition of what an "issue" is so that observers can spot its emergence. Nevertheless, as slippery as issue creation is, it is still the critical first step in politics. To understand why environmental disruption has been ignored and why governments have waited so long to act, we have to explore the factors delaying the transformation of pollution into an issue.

An issue is an "unresolved matter, controversial or noncontroversial, which awaits an authoritative decision." [1] The key words here are "unresolved" and "authoritative." An issue exists when a condition requires resolution; if it is not resolved, some facet of society will be in a state of uncertainty, conflict, or instability. If neglected, an issue can undermine political confidence. "Authoritative" refers to the *quality* of the resolution. There are scores of unresolved matters in any society on a given day. While these questions remain open, there will be a degree of uncertainty in the lives of those concerned. Most decisions made on any given day will be private choices. But "authoritative" resolution suggests that a decision will have to be made by some public official invested with governmental authority. So long as air pollution simply requires a decision by a private citizen it is not, strictly speaking, an issue. This is why anti-litter campaigns are so suspect to a politically committed environmentalist. By encouraging citizens to think that the environment can be preserved if only each citizen will make the private decision to refrain from throwing a

1. Matthew A. Crenson, *The Un-Politics of Air Pollution* (Baltimore: Johns Hopkins Press, 1971), p. 29.

bottle or a candy wrapper, the anti-litter campaigners are delaying the day when citizens realize environmental preservation requires hard choices by them collectively through their governments.

## CULTURES BENEATH THE ISSUES

Issues arise out of political culture and ideology. Values, expectations, and collective goals determine not simply how particular issues are resolved, but what aspects of public life are deemed serious enough to warrant governmental attention. What a citizenry deems "serious" and falling within the bailiwick of government are judgments derived from a political culture. The leader of a would-be movement must get other people to take the movement's issue seriously. When the environment appears to gain issueness quickly in a country, it is a sign that the country's culture includes ideas which focus attention on nature's delicate balance and treat it as a matter worthy of authoritative action.

Given the deep-seated character of cultures and given their uncanny adaptiveness in the face of new conditions, spokesmen for a new public concern may gain issue status for that concern by showing potential constituents how this subject fits neatly within their established mode of thinking and behaving. For instance, Swedes and Germans already possess collective norms that promote issueness for environmental affairs. By contrast, Americans' traditional outlooks obstruct such promotion. As a result, the achievement of issue status for the environment in Sweden and Germany may "cost" less culturally than it will in the United States.

A study of the contemporary Swedish society reveals the following values: (1) empiricism, (2) malleability of institutions, (3) legalism, (4) science and the usefulness of

knowledge, (5) privacy, (6) proper performance, (7) abhorrence of violence, (8) quality, (9) aesthetic sensitivity,
(10) nature, (11) international interdependency.[2]
With the possible exception of value 5, privacy, the
Swedish values enhance the likelihood of environmental
concerns being taken seriously in the political arena.
First, a society like Sweden, which highly regards science
and empirical verification, is likely to detect environmental problems sooner and take their effects more seriously.
Similarly, Swedes' belief in the malleability of institutions
encourages them to make organizational adjustments to
cope with new problems. Third, though foreigners in
Sweden frequently are puzzled by its citizens' reserve, it
fosters public civility and the performance of public roles
without undue emotion or subjectivity. Civility prompts
Swedes to regard industrialists' dumping or real estate
developers' land exploitation as a community concern
rather than a natural exercise of entrepreneurial initiative.

Perhaps the value best known abroad is the Swedish
love of nature, which takes the form of aesthetic pleasure
in cultivated landscape. The countryside dotted with
small farms cleared out of northern forests has been the
object of Swedish traditional identity. But as more and
more small farms are abandoned by city-directed mi-
grants, the forests gain ground. Swedes are reacting to
save their farming countryside, but the reaction is not
without its ambiguities. While some branches of the
Swedish government are now taking steps to preserve or
even roll back the forests by encouraging farmers not to
abandon their lands, other Swedish departments are
thinking first of economic growth and encouraging the
urbanization of subsistence farmers.[3] This warns us that

2. Richard F. Tomasson, *Sweden: Prototype of Modern Society*
(New York: Random House, 1970), pp. 273–82.
3. David R. Klein, "Cultural Influences on Landscape Aesthetics:

no culture is so simple and unambiguous as to guarantee a political life free from debate or inconsistency. A society's political culture determines what a people value highly or treat indifferently. It also shapes the processes by which they convert concerns into issues and then go about resolving them. Here again the Swedish case is instructive. Swedish values downplay mobilized public expression and give professional planners great leverage. Likewise, Swedish egalitarianism has meant Socialist party regimes in power. This combination of (1) eschewance of public demonstration, (2) widespread consensus on welfare programs, and (3) faith in professional planners gave Sweden a head start over other nations in dealing with environmental problems. This same political culture, however, also reduces conflict surrounding environmental matters to the point that an outsider might conclude that the environment was not a genuine issue in Sweden. An American urban planner studying in Stockholm reported that, unlike the United States, where "demonstrations, picket lines, leaflet distribution and other techniques which attract attention from the media" were commonplace in the planning process, in Sweden "such tactics were almost unheard of." [4] In Stockholm, "model city for the world's urban planners," as well as throughout the Swedish political system, "direct citizen participation in planning decisions was rare, aside from an occasional local flare-up." [5]

---

Some Comparisons Between Scandinavia and Northwestern North America," *Environmental Affairs* 2, no. 1 (Spring 1972): 82.

4. Shirley S. Passow, "Stockholm's Planners Discover People Power," *Journal of the American Institute of Planners* 39, no. 1 (January 1973): 23.

5. Ibid. Passow notes, however, that Swedish political culture may be changing to permit more conflict and public participation. During the so-called Battle of the Elms in Stockholm in 1972, planners and bureaucrats were forced to take account of protesting citizens before chopping down trees to make way for a transit system. At least one observer, however, doubts whether planning agencies ever were so powerful in Stockholm politics. See Thomas J. Anton, *Governing Greater Stockholm* (Berkeley: University of California Press, 1974).

Issues are determined not only by cultural patterns but also by ideology. Although culture is embedded in habit and presumption, ideology is calculated and explicit. Like a cultural pattern, an ideology includes norms and expectations that mutually reinforce one another, though more directly tied to action.

A new political issue may alert citizens to assumptions of their now-familiar ideology. Marxism and the American brand of democracy both represent ideologies that have found wide acceptance. Their differences are best known because they have fueled partisan and international hostilities since the end of the 1800s. The industrial revolution sharpened the differences between Marxism and democracy. The government's proper role in settling differences between factory owners and workers, the real nature of workers' "best interests," the effectiveness of "tinkering" as opposed to radical all-encompassing change in order to remedy the disparities born of industrial expansion—all these questions formed the issue-net out of which the Bolshevik revolution grew and the cold war was sustained.

By contrast, the environmental movement makes the differences between the two ideologies pale. Both American and Soviet elites hesitate to acknowledge pollution as a serious issue. This common reluctance suggests that, at bottom, Marxism and democratic theory share more than their respective proponents would care to admit. Each presumes economic and technological progress is natural and positive. "Progress" was an inherent good in the era of the industrial revolution, when both ideologies blossomed. What Marxists and democrats debated was not the desirability of progress but the proper *distribution* of its fruits. When Marxist and liberal democrat confront what to them appears to be the "fatalist" cultures of African and Asian societies, both are appalled. Technical emissaries of both went abroad selling the wonders of industrialization to the agrarian, underdeveloped world. Karl Marx might have mused about the joys of spending his postrevolutionary days in amiable conversation and going fishing; likewise, Thomas Jefferson might have delighted in his rural Virginia hilltop. But both authored ideologies that teach men to pursue and exploit technology and to reorder societies in such a way that exploitation can proceed most effectively. To take environmental issues seriously risked for both a "turning back of the clock" and a retarding of progress.

China illustrates how this pro-material growth, pro-industrialization bias of the nineteenth-century ideology can be turned around to promote, not discourage, serious concern over environmental issues. The Chinese Communist elite believes in progress: "Chairman Mao has taught us: 'In the fields of the struggle for production and scientific experiment mankind makes constant progress and nature undergoes constant change; they never remain at the same level. Therefore, man has constantly to sum up experience and go on discovering, inventing, creating and advancing. Ideas of stagnation, pessimism, inertia and complacency are all wrong.' " [6] Yet what allows environmental concerns access into the Chinese political arena is Mao's concept of "Contradictions." For Mao and his colleagues there is no permanent resolution; one must constantly monitor change, be on the lookout for new conflicts, and derive social benefits from those conflicts.

When interpreted in the light of the theory of contradictions, progress and economic development seem more mixed in their benefits and costs. For instance, while Russian theorists are inclined to dismiss pollution as a capitalist malady, the Chinese acknowledge that, although capitalism's reliance on profit as the sole criterion

6. Fang Hsin, "Economic Development and Environmental Protection," *Peking Review*, no. 29 (20 July 1973): 8.

for measuring success is a major cause of environmental deterioration, pollution can be a problem even in socialist countries. Pollution comes not just from the ownership of production but from the very character of production itself. Speaking in Maoist terms, if environmental costs are figured into accounting and political planning, then they will remain "unantagonistic contradictions." If those costs are ignored for the sake of unimpeded economic growth, however, then an unantagonistic contradiction may eventually turn into an "antagonistic contradiction" —that is, it will subvert the revolution.

Ecology was not a prominent element in the Chinese Communist ideological formation. Only in the later 1960s and especially with its participation at the 1972 Stockholm Conference has the Chinese regime spoken out on environmental questions. One reason for the delayed emergence of environmental issues in Chinese politics, of course, is the relatively low level of industrial development. Shanghai, Canton, and Mukden were heavily industrialized urban centers; but the overwhelming proportion of China's territory and citizenry were rural. In addition, the very assumption that Maoists made about the harmful effects of production deriving solely from the economic distortions imposed on it by capitalism slowed the emergence of the pollution issue. Admitting that pollution was a problem requiring political resolution would be tantamount to admitting that vestiges of capitalism persisted under communist rule.

Perhaps, then, it is not mere coincidence that environmental questions took on greater political visibility in China during the Cultural Revolution. In other words, mounting public attention given to environmental disruption in China in the late 1960s was not merely a spinoff from the international ecology movement symbolized in the Stockholm Conference. Rather, new awareness may derive more from the sharpening "contradictions" in the

Cultural Revolution. The main adversaries in the Cultural Revolution, personages powerful in their own right but also made to symbolize much broader ideological and policy conflicts, were Mao Tse-tung and the CCP Secretary General Liu Shao-chi. Liu was accused of putting economic priorities above political priorities, leading China down the path of revisionism. He was referred to in the Chinese press not by name, but as "China's Khrushchev." To disregard the Maoist dictate of "Politics in command" and to be preoccupied with economic growth for its own sake meant to subjugate the human needs of the masses under abstract measurements of productive efficiency. It is just such a distortion that exacerbates the inherent contradiction between productivity and waste. Thus, out of a nationwide controversy about the proper goal of economic development and its relationship to other socialist goals came a greater sensitivity to the environment as a political issue.

Currently, three slogans define the ideological interpretation of pollution in China. Slogans are a shorthand version of ideological theory, condensed and simplified so that ideology can mobilize people into action. Not all slogans are ideological. "Keep America Beautiful" is not ideological. It does not plug into a wider network of slogans which, when taken together, define reality and set forth mutually compatible goals for society. But in China slogans for environmental actions are intensely ideological:

*"Everything divides into two."*
This is essentially a dialectical perception recognizing that "environmental problems which initially appear harmful can be transformed into something beneficial." [7]

---

7. J. B. R. Whitney, "Ecology and Environmental Control," in Michael Oksenberg, *China's Development Experience* (New York: American Academy of Political Science, 1973), p. 104.

*"Struggle against the four wastes—waste material,
waste gas, waste water and waste heat."*
Because waste is an ideological concern, not merely
an operational problem, to engage in waste is not just
inefficient, it is unethical.[8]

*"Walking on two legs"*
Traditional methods of operation are not in themselves bad, it is the way in which they were used
to oppress the peasants and proletariat that made
them objectionable. A large underdeveloped society
such as China, in order to ensure self-reliance and
widespread popular participation in national efforts,
must draw upon traditional techniques as well as
modern techniques. This means avoidance of dependency on mechanization for its own sake. Walking on
two legs thus fits in with the struggle against waste in
its sanctioning of the use of human waste in agriculture and not relying on chemical fertilizers.

An ideology links issues. Environmental control has
added momentum and saliency in China, now that it has
achieved issue status, because China is a highly ideolo-
gized society. Recycling in China relates to a vast range of
critical political decisions and values, while recycling in
an affluent American suburb stands as a somewhat
isolated social act. In fact, in the American upper-class
suburb, recycling may be enthusiastically supported by
the townspeople because it is nonideological. If taking
one's milk cartons and TV dinner aluminum trays to the
recycling centers were perceived as expressing some
explicitly political outlook, as it would be in China, many
suburbanites might avoid the center.

Ideological saliency is no guarantee of environmental
purity, however. As a Canadian journalist in China
reported, "On a clear day in Peking you cannot always

8. Orville Schell, "China's Way with Waste," *Ecologist* 3, no. 2
(February 1973): 56.

see forever."[9] Likewise, a Canadian steel executive described the air pollution in Anshan, Manchuria, where China's biggest steel mill is located, as the worst he had seen in his professional lifetime.[10] In international contexts, furthermore, Chinese still find it embarrassing to admit their shortcomings because they jeopardize Peking's role as leader of Third World nations.[11]

## INCIDENTS AND ISSUE CREATION

Culture and ideology have long-term effects on how individuals perceive the world around them and respond to it. But in the short run, concrete incidents stimulate interest, desire, or anxiety and alter the demands placed on governments. The interaction between collective values and shared events is the stuff out of which political issues are made.

An event provokes environmental issues when it (1) stimulates media attention; (2) involves some arm of the government; (3) demands governmental decision; (4) is not written off by the public as a "freak," one-time occurrence; (5) relates to the personal interests of a significant number of citizens. Among those incidents that have served to politicize the environment have been mercury poisoning of fish in Japan, the New York City "blackout," the wreck of the oil supertanker *Torrey Canyon* off Britain, and the London "killer smog."

Smoky air was not new to England. Nor were the health dangers that accompanied industrial, automotive,

9. John Burns, of the Toronto *Globe and Mail,* in *New York Times,* 19 August 1973.

10. Ibid.

11. Sources on ecology policy in China also include: Irving Stone, "The Ecoscene in China," *International Conservationist* (British Columbia) (Winter 1972/73): 8–10; "Quarterly Chronicle and Documentation: United Nations," *China Quarterly,* no. 51 (July/September 1972): 593–602; Leo A. Orleans and Richard P. Suttmeier, "The Mao Ethic and Environmental Quality," *Science,* 11 December 1970, pp. 1173–76.

and home-furnaces pollution. Until 1952, however, Britons considered smog the inevitable price paid for industrialization. In the 1800s Charles Dickens vainly attempted to arouse his fellow countrymen to the evils of industrial growth uninhibited by governmental regulation. In his satirical novel *Hard Times* Dickens has his leading spokesman for the benefits of the industrial revolution explain to a visitor to industrial Coketown:

> First of all, you see our smoke. That's meat and drink to us. It's the healthiest thing in the world in all respects, and particularly for the lungs.[12]

Britain's Midlands residents contend, "Muck is money." Smoke-filled air was not a notable incident; it was a profitable way of life. Americans shared the same outlook. Not intending to be satirical, Republican presidential candidate Wendell Willkie told a campaign rally in the milltown of Lowell, Massachusetts, at the end of the Great Depression that their smokeless factories were a symbol of the New Deal. The smokeless air, reminded Willkie, was a sign of the harm Roosevelt's policies were causing working people. The "punitive, restrictive laws" of the Democrats had "made it impossible for the United States to recover quickly from the depression." [13]

Thus it becomes valuable to analyze the London smog of 1952 in order to discover what incident it takes to transform public tolerance into political intolerance. The damp, chilly smog lasted for an entire week in December; when it lifted, four thousand people were dead and thousands more were left ill. The result of the smog was both the birth of an issue and the passage of one of the West's first major pieces of pollution legislation, Britain's

---

12. Charles Dickens, *Hard Times* (New York: E. P. Dutton, 1907), pp. 112–13.
13. *New London Day* (Connecticut), 12 October 1940.

Clean Air Act of 1956. On the other hand, there is evidence that since the mid-1950s the air pollution issue has declined markedly in Britain, suggesting that issues can wane as well as wax and that the mental impact of a particular incident may be short-lived.

There has been a Coal Smoke Abatement Society in Britain since the nineteenth century. One of its founders was Harold Des Voeux, the first person to coin the term "smog." [14] Despite the Society's efforts, air pollution did not capture either the public's or the politicians' attention. Its failure followed in the train of failures of earlier issue makers. In the 1680s London pamphleteer John Evelyn tried to alert the Stuart monarchs to the dangers arising from the burning of coal. In 1684 Evelyn described for probably the first time what is now termed a temperature inversion, the weather condition causing the killer smog of 1952. "London," Evelyn wrote in January 1684, "by reason of the excessive coldness of the air hindering the ascent of the smoke, was so filled with the fuliginous steam of the sea coal, that one could hardly see across the streets, and this filling the lungs with its gross particles, exceedingly obstructed the breast, so as one could hardly breathe." [15] Almost three hundred years later, only limited and piecemeal governmental actions had been taken to control air pollution in the London area. Furthermore, there was little public pressure for government to do more. Thus in the autumn of 1952, just several months prior to the historic smog, the Conservative Minister of Housing and Local Government could reply to an opposition Member of Parliament that he was satisfied with the

---

14. L. T. Foster, "The Development of the Smokeless Zone Concept," in *Papers on Selected Social Aspects of Air Pollution in the United Kingdom* (International Geographical Union, Commission on Man and Environment, Calgary Symposium, 23–31 July 1972), p. 2.

15. William Wise, *Killer Smog* (New York: Ballantine Books, 1970), p. 20. The following description of the 1952 London smog is based on Wise's account.

current level of governmental pollution control and that he did "not think that further legislation is necessary at present." [16]

What makes the week-long air pollution disaster of 1952 so politically instructive is that it was not perceived as either a disaster or cause for political action while it was going on. Firsthand recountings of Londoners' experiences during the first week of December 1952 reveal that individuals did not know the harmful, and in many cases, fatal consequences of the smog. They were not alerted by any official authority to stay indoors; they did not look for culprits or remedies. One physician told a reporter afterward that when on the second day he began to suspect that London was experiencing a calamity akin to those of Donora, Pennsylvania, and the Meuse Valley in Belgium (neither of which produced a political issue), he tried to stimulate action in the Department of Health, but with no success. He tried to alert the public through the press. But no BBC news report issued warnings and no references to Donora or the Meuse were made.

Often in the midst of a calamity—a fire, riot, earthquake, power blackout—the broader implications for public well-being and government responsibility are overshadowed by immediately pressing demands. So it may not be surprising that during the smog itself neither British media nor government spoke in issue terms. Even after the air cleared, however, there was reluctance to perceive the week's temperature inversion as a matter to be coped with politically. First, data on the climatic nature of the incident were scarce. Second, even scarcer were data on the injurious effect of the incident. To become an issue the smog would have to be defined and its threat to public health specified. Sufferers from the dense smoke, turning groundward rather than skyward,

16. Ibid., p. 62.

were the very old and the very young—the members of society who are always most vulnerable. The incidence of *their* illnesses and fatalities during the days of smog was not perceived as extraordinary. To be an issue, the smog would have to be seen as causing some unnatural and thus avoidable illness among the ordinarily healthy.

When two M.P.s called for an official inquiry into the smog, the Conservative government rejected the proposal. Then, in January 1953, the highly respected *British Medical Journal* published a report that as many as 4,700 deaths had been caused by the smog. Later in January the London County Council issued its annual medical report stating that the smog had probably been more deadly than the worst nineteenth-century cholera epidemic. It was this new information from respected professional sources that equipped M.P.s with the means for compelling the Cabinet to take the smog incident seriously as an issue. Still, the ministers delayed. There were no votes to be gained from scouring the atmosphere. Only when the concrete figures of illnesses and fatalities, along with specific comparisons to other air pollution crises, were published could organizations such as the National Smoke Abatement Society (descendent of the Coal Smoke Abatement Society) mobilize enough public opinion to prod the government out of its inaction. The Society conducted its own survey of the quality of air and extent of human and economic damage. It circulated the surveys' findings to all Members of Parliament. M.P.s began to receive mail from constituents demanding official enquiry.

Finally, in the late spring of 1953, the British government acknowledged that an issue existed. It established an Air Pollution Committee to conduct an official investigation of the smog. The committee's chairman, Sir Hugh Beaver, believed that if the public were without informational stimulation during the duration of the investigation the issue might die, as many officials hoped it would, and so he released an Interim Report at the end of 1953, close to the anniversary of the killer smog. The Interim Report found both the national government and local authorities "guilty of negligence in failing to take all possible steps to protect the people of London from a smog disaster," at the same time laying partial responsibility at the door-steps of average Londoners, who continued to burn coal in their furnaces.[17] This report not only kept the issue alive but also made the political dimension of the smog explicit. The Final Report of the Beaver Committee was issued in 1954 and led to Britain's first effective environ-mental legislation, the Clean Air Act.

An "incident," essentially, is the intensification of reality to such a degree that it becomes visible and concrete to the witness. Scientists can make ordinarily routine, invisible phenomena visible by the collection of data demonstrating the significance of environmental phenomena for public well-being. When Harold Des Voeux gave us the word "smog" he fashioned a tool with which to distinguish incidents and thus to stimulate issues. In education, welfare, arms control, and trade, issue making is in large part the definition of terms and the specification of consequences. So long as Londoners and their M.P.s could shrug off the smog as unremarkable and having no clear-cut harmful results, it was possible for air pollution to remain outside the issue arena.[18]

To create an issue may require the creation of a new profession or the raising of an existing profession's public status. A Hungarian sociologist underscored this point

17. Ibid., p. 198. Also see Foster, "The Development of the Smokeless Zone Concept," p. 10.

18. A less publicized but equally instructive case of the importance of measurement in creating political issues is the controversy following the oil spill in the Cape Cod Canal, September 15, 1969. See William Wertenbaker, "Reporter at Large: A Small Spill," New Yorker, 26 November 1973, pp. 48–79.

when he urged his colleagues in Eastern Europe to redistribute professional rewards so that more weight would be given environmental data and more skilled persons would enter environmental professions:

> To put it clearly: even creating a Ministry of Environmental Protection and putting in prison those who contribute beyond measure to the pollution of water and air is of no avail (so long as) e.g., a young forester . . . is less likely to win the hand of the prettiest girl in the village than a factory foreman from the next town.[19]

## INDICATORS OF ISSUENESS

Environmental disruption has existed in many countries for generations, yet only within the last two decades has it turned up as an issue demanding political choice. Sweden and West Germany have been among the vanguard; France has been reluctant and China often ambivalent; Guyana considered the environment such a nonissue that its government did not bother to send delegates to the UN Conference on Man and the Environment held in Stockholm in 1972.

To determine whether environmental pollution has become a political issue one needs indicators of "issueness." What conditions signify that the environment has aroused unresolved questions requiring government decision making? A list of indicators might include:

1. Media. Do newspapers and broadcast networks

---

19. Andras Szesztay, "East Central European Models of Environment Protection" (Paper prepared for the Budapest Regional Conference of the International Geographic Union, Budapest, August 1971), p. 12. A survey of American college youth in 1973 found that "environmental protection, city planning and social work are popular career choices." "Children of the Revolution," *Fortune*, March 1973, p. 158.

assign reporters to cover environmental stories on a regular basis?

2. Elections. How often do environmental issues dominate candidates' campaigns? If the system allows for voter referenda, are environmental questions on the ballot and is voter turnout high?

3. Legislatures. Do representatives seek positions on committees dealing with natural resource policy? Or do they just leave those posts to a handful of representatives with special interests?

4. Organizations. Do nongovernmental groups organize around environmental causes and are they given access to policy makers?

5. Political parties. Are parties including environmental planks in their platforms?

6. General groups. How often do interest groups with diverse goals—labor unions, scientific associations, conferences of municipal officials—debate environmental policies?

7. Polls. Do public-opinion pollsters include questions on environmental affairs? Do the polls show a public awareness of government policies affecting the environment?

8. Business. Whether or not they actually act to conserve the environment, do business firms buy advertising space to express their concern for the environment?

9. Officials. Do powerful officials justify policies, regardless of whether the justification or reassurances are valid, in terms of their impact on the country's environment?

10. Power. Can stands on an environmental issue cause a major politician to rise or fall from national power?

Clearly, the same indicators of issueness do not hold for all political systems. The utility of each indicator has to be assessed in the context of the particular political system. Probably the tenth indicator has the most widespread applicability. In countries as dissimilar as the Soviet Union, Canada, and Nigeria there are political officials who possess disproportionate amounts of power and are key articulators of governmental policy and its rationale.

Coverage of environmental affairs in the press and on radio is a more universal indicator today than it was sixty years ago. Still, the role of the media can differ sharply from nation to nation, so the use of the media indicator of issueness should be adjusted accordingly. For instance, the UN Stockholm Conference had press coverage on the editorial pages in Guyana in 1972 even though the government declined to send Guyanese delegates. On closer examination this discrepancy is less puzzling, for Guyana's major newspapers rely on foreign wire services for coverage of overseas events. Thus, while Guyanese newspaper readers were made aware of this important international conference, press coverage was a reflection of Reuter's perceptions more than it was an indicator of the priorities of local Guyanese editors.

A survey conducted in the United States in 1970 sought to determine how much space American newspapers devoted to environmental affairs and how important the editors expected environmental news would be during the 1970s. The sample of newspapers chosen by the researchers was drawn from a federal government list of *high* air pollution cities. Thus the survey results reflect levels of media concern in places with the greatest objective need for issue awareness. More than half of the newspaper editors in those cities thought that pollution would be a major news story in the 1970s. On the other

hand, when asked their reasons for making this predic-
tion, the belief that pollution would be a "big political
issue" was mentioned far less often than the belief that
there was already public interest and that pollution
caused genuine harm.[20]
    The editors' predictions can be self-fulfilling. If they
believe pollution will be a major story in the forthcoming
decade they will want to be up-to-date and thus plan for
coverage. This planning will then ensure that pollution is
a regular news beat and gets press space, which in turn
will keep it before the public eye and generate continuing
interest. Likewise, however, if relatively few editors
believe that the public's interest in pollution has a
political issue basis, then pollution reporting is less likely
to be couched in political terms and more likely to be
relegated to the paper's science, business, or leisure
sections, thus retarding readers' perception of pollution
as a matter rightfully under government control. Already
in 1970, twenty-two of the large-city papers which an-
swered the questionnaire noted that they had reporters
regularly assigned to the pollution beat, eight on a
full-time basis. In just a single month, January 1970, the
sampled papers reported that they each devoted an
average of 8.4 page-one stories and 4.5 editorials to
pollution.[21] Nevertheless, it may be up to interest groups
and politicians to push these stories from the general-
interest columns to the political columns.
    Media analysts warn, however, that while the amount
of media coverage affects public *awareness* of an issue, it
does not determine the specific *attitudes* that citizens
hold on that issue.[22]

---

    20. John C. Maloney and Lynn Slovonsky, "The Pollution Issue: A
Survey of Editorial Judgements," in *The Politics of Ecosuicide*, ed. Leslie
L. Roos, Jr. (New York: Holt, Rinehart & Winston, 1971), p. 68.
    21. Ibid., p. 72.
    22. G. Roy Funhouser, "The Issues of the Sixties," *Public Opinion
Quarterly* 37, no. 1 (Spring 1973): 74.

Issue creation is a process that occurs over time. Thus even more revealing than this snapshot portrait of press coverage of pollution is a study showing the pattern of rising or falling coverage over several years. A survey of articles listed in the *Guide to Periodical Literature* from 1953–69 produces a graph climbing steadily upward, indicating increasing U.S. magazine coverage of environmental pollution. Again, however, such an increase cannot be read as an indicator of political issue awareness since there are several contexts in which a reporter can place air or water pollution, many of them devoid of political implication. The study shows that, while in 1953 there were a mere 50 articles focusing on environmental issues in these magazines, by 1969 the figure had risen to almost 250.[23] The survey also revealed a changing emphasis within the broad area of environmental affairs. Between 1953 and 1969 American magazine writers directed more and more of their attention to *urban*-related environmental matters. Likewise, the problems resulting from industrialization and specifically the serious character of air and water pollution sharply increased in coverage.[24]

We have few equivalent studies from other countries. Nevertheless, rough indicators suggest a similar rise in media coverage of environmental affairs. First, several American publications—*Newsweek, Time, Reader's Digest*—have international circulations and carry their new environmental orientations to readers in Asia, the Caribbean and elsewhere. France's environmental issueness is lower than in most European countries. But in France now there is a weekly television program devoted to the environment, "La France de figuree." Also, *L'Express*, the French weekly newsmagazine published by one of

---

23. James McEvoy, "The American Concern with Environment," in *Social Behavior, Natural Resources and the Environment*, ed. William R. Burch et al. (New York: Harper & Row, 1972), p. 218.
24. Ibid., pp. 219–20.

France's best-known political spokesmen, Jean-Jacques Servan-Schreiber, initiated a new section devoted to the environment in the late 1960s. It has covered such environmental debates as those surrounding the French-British supersonic plane, the Concorde, and the French explosion of nuclear devices in the Pacific—both set squarely within a political context.[25]

The devotion of media time to the environment may be an especially potent indicator if the media manage to enlist the support of celebrated public figures. Citizens of the Netherlands, for example, have had to be sensitive to environmental relationships for centuries. The country exists in large part as a result of man's ability to control seawater levels. But protective, as versus exploitive environmental, concerns are relatively new in Dutch politics; and the media have played a role in giving the environment its issue status. Public figures and politicians appeared on television and radio to highlight environmental problems. Prince Claus, the husband of Crown Princess Beatrix, appeared in a television series to advocate environmental reform. The 1970–71 series was entitled "We Stinken Er In" ("We Are Trapped into It"). These royally sanctioned television shows stimulated other media. "The reaction to these broadcasts was extensive and positive; they were thoroughly covered by all of the country's newspapers and magazines, and were given overwhelming approval in the numerous letters sent to the producer and the broadcasting network."[26]

Besides media coverage, one can look at political contests to estimate issueness. To what extent does the environmental issue cause those in power to rise or fall?

---

25. After the French presidential elections of 1974, Servan-Schreiber was appointed Minister for Reform in the new government, but resigned in protest when President Válery Giscard D'Estaing insisted on one final atomic test.

26. Manfred C. Vernon, "The Netherlands and Its Waters," *Environmental Affairs* 2, no. 1 (Spring 1972): 206.

While more parties, especially in Western Europe, the United States, and Canada, recognize environmental concerns in their party platforms, no national elected leader has gained or lost national power because of his or her stand on the environment. Italian and Dutch difficulties in forming stable governments have not been caused by popular or partisan conflicts over environmental issues, though in both cases the critical economic conflicts relate directly to environmental questions. In Canada, the Liberal party and Pierre Trudeau's loss of parliamentary confidence in 1974 cannot be laid at the feet of environmental disputants; inflation, U.S. hegemony, and French Canadian nationalism were far more crucial to this decline in party power. Lyndon Johnson did not decline to run in 1968 for the U.S. Presidency because of inability to cope with environmental issues; neither Richard Nixon's electoral victories in 1968 and 1972 nor his resignation in 1974 can be traced to popular environmental demands. No Chinese or Russian Politburo member has been purged because of his stand on water or air pollution. The success or failure of agricultural harvests, attitudes toward national security, factional allegiances to other political figures—all have been stronger determinants of Soviet and Chinese political careers. Only in Japan can one say that pollution has become such a significant political issue at the national level that it has affected the careers of major elected officials. Nevertheless, although pollution is a primary political issue in Japan, the party that has governed Japan during the rise in pollution from 1955–70 remains in power today.

The issueness of environmental pollution appears more influential in political careers at the local level than at the national level in most countries. It has been mayors, state and provincial officials, and national representatives of local districts that have risen and fallen on the waves of politicized environmental concern. In Canada's western province of British Columbia, for example, the 1972 election brought victory to the New Democratic party under the leadership of David Barrett when he campaigned against encumbent Premier W. A. C. Bennett's exploitation of the province's natural resources.[27] Again at the local level, a three-way mayoralty contest in Toronto in 1972 was fought over the merits of homeowning neighborhoods versus developers. The candidate with the strongest stand in opposition to development harmful to the urban environment, David Crombie, won Toronto's mayoralty in an upset victory. Press commentators credited the environmental issue giving Crombie his win.[28] In the Netherlands too, the environmental issue had its greatest electoral potency at the local level. In the early 1970s a small upstart party called the Gnomes won 5 of the 45 seats on the Amsterdam City Council. The Gnomes ran on a platform giving primacy to environmental recommendations. More important than their council seats, perhaps, is the fact that senior government officials considered the radical leaders of the Gnome party to be prime movers of Dutch public opinion.[29]

Why should the issueness of the environment appear most forcefully at the subnational level? Why should the public and official sense of the need for resolution and authoritative decision emerge earlier in city or provincial politics than in national politics? First, the very character of environmental discomfort is localized. While international trade imbalances or taxation usually are felt in a generalized, nationwide manner, smoggy air or strip-mined landscape is experienced in a geographically immediate way. Secondly, in many political systems responsibility for public health, sanitation, and zoning standards reside with local and provincial or state officials.

27. Allen Garr and Bob Waller, "The Pacific Persuasion," *McLean's* (Canada), June 1973, pp. 22–32.
28. *New York Times*, 2 December 1972, 15 December 1972; *Boston Globe*, 17 December 1972.
29. *New York Times*, 6 July 1973.

In the future local officials and residents may recognize that environmental problems are nationwide and that the authority and money needed to resolve these problems lies only within the central government. When such wider perspectives develop, the environment's issueness may become as critical nationally as it is locally, to the point that prime ministers and presidents will rise and fall on their environmental stands. In the mid-1970s that is not yet the case.

Whether at the national or subnational level, the indicator for issueness is not the victoriousness of pro-environment candidates. Even had the NDP lost in British Columbia or the Gnomes lost in Amsterdam, the index of issueness would have registered high. For in each the environment was seen as a key question on which voters evaluated alternatives.

Two other indicators of issueness need special attention if they are to be used comparatively. A referendum is a valuable indicator because it leaves less to the imagination of the onlooker, since the issue itself is on the ballot. Oregon has had a referendum on no-return bottles; California voters faced a referendum on preservation of coastal lands; Colorado held a referendum asking whether residents wanted to host the Winter Olympics, an event charged with having serious environmental repercussions.[30] In other countries referendum elections are held nationwide. One of the few national referenda that was fought in part on environmental questions was the Norwegian referendum of 1973 on entry into the European Common Market. Norwegians voted against entry, and one of the rejoicing nay-sayers was a group

---

30. In Switzerland a local referendum was fought in Zurich to prevent underground transit lines. "Switzerland: View from Below," *Economist*, 26 May 1973, p. 48. And in Yugoslavia, one town's citizens decided via a referendum to dock their own salaries in order to finance the cleaning of their local lake which drew tourist business to the region. *New York Times*, 23 July 1973.

contending that entry would spur industrialization of Norway and harm its environment.

The study of referendum voting is a neglected field. Most of our knowledge of electoral behavior in the United States and abroad has derived from the more common candidate-selection balloting. The frequency with which environmental decisions are being made in public polling booths suggests that we need more investigations of referenda behavior.

One other indicator of issueness deserves brief discussion. A subject becomes a political issue when political spokesmen acknowledge it to be salient to their constituents. Now, a politician can refer to pollution frequently without effectively sponsoring environmental bills. His references may take the form of belittling remarks about "enemies of progress" or "unrealistic reformers." But the fact that a person whose influence depends on public support acknowledges that environmental concerns have moved from the private, noncontroversial realm into the arena for authoritative resolution is indicative of issueness.

*McLeans*, one of Canada's most influential magazines, recently sent questionnaires to the ten Canadian provincial premiers. In the Canadian system the premiers carry notable political power because of decentralized federal authority and province-based political party organizations. Thus, what the provincial chiefs label important can provoke considerable interest. Since the late 1960s environmental disruption has elicited growing attention in Canada, a country still possessing frontier regions and a relatively small population.[31] *McLeans'* poll of the ten provincial premiers suggests that other issues

---

31. In *McLeans* itself there was a rise in environmentally related articles: 1971, three articles; 1972, three articles; 1973, six articles within the first six months. The specific issue that had greatest coverage was the northern oil pipeline, a question that touches the increasingly raw nerve of U.S.–Canadian relations.

remain far more prevalent and salient than the environment. Each premier was asked what he considered to be "the three most serious problems facing his province today." In addition, each was asked what three issues he considered the most serious for Canada as a whole. Here are the premiers' replies:[32]

|                      |            |                                                             |
|----------------------|------------|-------------------------------------------------------------|
| *Newfoundland:*      | National:  | regionalism; economy and unemployment; preserving the environment. |
|                      | Provincial:| lack of natural resources; conservation; unemployment.      |
| *Prince Edward Island:* | National: | lack of investment; small population; rising costs of social services. |
|                      | Provincial:| lack of natural resources; lack of investment on secondary industries. (Only two issues stated.) |
| *Nova Scotia:*       | National:  | regional disparity; national unity; world trade.            |
|                      | Provincial:| costs of government services; unemployment; national policies incompatible with the province. |
| *New Brunswick:*     | National:  | federalism; unequality of opportunity; tend-                |

32. "Politics of Taste," *McLeans* (Canada) 86, no. 2 (February 1973): 19–25.

ency to oversimplify
problems.

Provincial: unemployment; lack of
government revenues.
(Only two issues
stated.)

*Ontario:*       National: unemployment; taxation and economy;
environment.

Provincial: Same as national.

*Quebec:*        National: "identity crisis"; excessive centralization;
fiscal redistribution.

Provincial: unemployment; federal
centralization; fiscal
redistribution.

*Manitoba:*      National: agricultural decline
and western alienation;
regional disparity;
foreign ownership and
unemployment.

Provincial: agricultural decline
and western alienation;
resources development;
native job opportuni-
ties.

*Saskatchewan:*  National: threat of absorption
culturally, economi-
cally, and eventually
politically by the
United States; regional
conflict; unemploy-
ment.

Provincial: role of agriculture,
especially its industri-
alization; reduction of

|              |             | reliance on agriculture; unemployment. |
|--------------|-------------|----------------------------------------|
| *Alberta:*   | National:   | job creation; national unity; world trade. |
|              | Provincial: | overdependency on natural resources. (No other reply.) |
| *British Columbia:* | National: | "lack of control of our economic and social destiny." (No other reply.) |

The environment is mentioned only twice by the Canadian premiers among top-priority national issues. It is referred to twice among provincial-level issues. Far more salient, clearly, are worries about unemployment, regional conflict (including Quebec separatism), and U.S. domination. But if one looks closely, many of the issues noted in the questionnaire are intimately connected to the environment, especially U.S. domination, which takes the form of polluting and extractive industrial investment, natural resource exploitation, or underindustrialization. Furthermore, the environment is mentioned by premiers from different parts of Canada. Whereas *McLeans'* own articles over the last few years highlight western and northern environmental issues, the issueness of the environment is evident in the eastern provinces of Canada as well.

Each of these trial attempts at testing environmental awareness strongly suggests that while indicators of issueness should be helpful in determining whether environmental affairs have gained the status of an issue, each indicator has to be utilized with a sensitivity not only to the differences among national political systems, but also to the political differences *within* a given country.

## ISSUE WEBS AND CYCLES

Issues rarely exist in a vacuum. Rivalries for scarce resources, multiple interests possessed by the same group, cultural and ideological linkages between values— each serves to attach one issue to several others. The web of issues into which pollution falls will influence the way in which it is debated and the kinds of social cleavages it provokes. This issue web may differ from country to country.

At first glance it benefits environmental advocates to be part of such a web. Linkages between pollution and, say, Indian rights or regional decentralization should aid in expanding political alliances and attracting media coverage. Frequently, it is the ideologue who is best equipped to see these linkages because ideology itself encourages the drawing of relationships between what otherwise appear to be random or fragmented conditions. The ideological spokesman can then map out alliances among diverse interests because he can show each the common root of their plights.

On the other hand, the web can make it difficult to isolate specifically environmental questions. The result will be that, instead of deciding simply on whether or not a chemical plant should be built beside Russia's great Lake Baikal, the question will be inflated into whether or not the Soviet Union can afford to relax its postrevolution drive toward economic modernization. In such a debate Lake Baikal's water quality is unlikely to benefit.

Figure 1.1 suggests some of the contrasting issue configurations in which pollution and environmental disruption is embedded.[33]

---

33. For details of Israel's belated and reluctant acceptance of the environmental issue, see Amos Elon, "Israel at Twenty-Five," *New York Times Magazine*, 6 May 1973, pp. 33–82; *New York Times*, 7 March 1973, 26 March 1973.

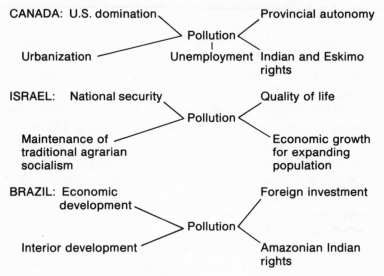

*Figure 1.1.*  **Issue Webs**

These issue webs are not balanced in a stable equilibrium. A public's concern for one issue can grow, leaving little time and energy to worry about the others. Thus it is the dynamic relationships between issues in the web and issues just "off-stage" that determine the seriousness and urgency with which citizens consider any single issue.

The very imbalance in a country's issue web implies that issues not only emerge, they also decline. The shallowness of an issue like environmental control may become apparent when it is challenged by a more potent issue. Environmentalists and their critics in industrialized nations both noted the readiness with which voters cast off environmental worries in the winter of 1973 as they became preoccupied with the availability of heating oil, gasoline, fertilizer, and petrochemicals. Suddenly, environmental issues were treated as if they were icing on the political cake—interesting to think about so long as the "essentials" of society were assured.

There are at least three possible causes for issue decline. First, an issue may fade in the web because it has been resolved. Child labor, slavery, church control of state, all have been primary issues which now have waned in many (not all) countries because means were found to resolve the conflicts. If technology improves, government regulation is expanded and social ethics becomes less materialistic, then the environment too may become a nonissue. But the complexity and universality of this issue makes decline via resolution unlikely for environmental questions.

Alternatively, then, an issue may fade because it is overshadowed by another issue that the public considers more pressing, more threatening, easier to comprehend in its implications for the ordinary individual. If this issue is one that not only overshadows the current issue but actually challenges its very legitimacy, the decline may be politically deadly. This is why environmentalists were so dismayed at public and governmental willingness to turn attention to the energy "crisis"—resolving the energy crisis appeared to require denying the *legitimacy* of environmental concerns. Famines could have the same effect on the environmental issue's life span in underdeveloped countries.

Third, issues may wane simply because most people have short attention spans. Political economist Anthony Downs has postulated a five-stage issue cycle.[34]

1. "The pre-problem state"—when a condition exists but without mobilized public concern.
2. "Alarmed discovery and euphoric enthusiasm"— public arousal and strong desire to do something

---

34. Anthony Downs, "Up and Down with Ecology—the 'Issue-Attention Cycle,' " *Public Interest*, no. 28 (Summer 1972): 38–50.

to solve the problem, with the accompanying assumption that the problem has a "solution."

3. "Realizing the cost of significant progress"—the public comes to recognize that the solution will require some sacrifices, even if only higher taxes. Faith in technology may lull the public for some time, however, into thinking the solution is "free."

4. "Gradual decline of intense public interest"—a growing realization of how difficult any solutions will be dampens public commitment and eventually even interest in the problem. "The poor will always be with us" mentality takes over.

5. "The post-problem stage"—the issue moves into limbo where it is of concern to a handful of specialists and makes headlines only sporadically, though now with newly created institutions or specialists to keep it at least partially alive.

Downs concludes that while all issues are subject to this cyclical pattern, environmental issues may be especially so because of their technical complexity, the costs required for their solution, and the breadth of the problem which seems to defy any easy solution.

It may be, as some Europeans have said, that Americans more than most people suffer from extremely short attention spans. It may be also that Americans more than other people doom an issue at the start by presuming easy "solution" and limited social sacrifice. The patterns of emergence and decline of issues will vary from country to country not only because the environmental question will find itself involved in varying issue webs, but also because different political cultures conceive of public problems and time frames of resolution differently.

## CONCLUSION

Before leaving the impression that issue creation depends only on ideology, culture, incident, and measurement, we need to hoist a warning flag. It is all too easy to assume that issues will emerge automatically whenever citizens are sufficiently exercised to express concern politically. Some social scientists warned that this model of issue making is too simple. Specifically, it ignores the intrusion of *power* into the process that transforms objective conditions into issues. Critics of the pluralist model of American politics, for instance, find power can be wielded not only in such a way as to obtain a policy favorable to a particular interest group, but also in such a way as to ensure that a problem never becomes an issue at all.

Matthew A. Crenson, in his provocative study of air pollution control in East Chicago, Illinois, and Gary, Indiana, concludes that the different timings of the emergence of the air pollution issue stemmed from the different power positions of the steel companies in the two cities.[35] East Chicago viewed air pollution as a political issue seven years before Gary did. Why? Imaginative comparative analysis allows Crenson not only to answer this question but also to shed light on the impact of power inequities on the process of issue creation. Although East Chicago and Gary are cities much alike economically, politically, and demographically, they differ from one another in the character of their respective steel companies' political roles. In East Chicago there were several steel companies. These companies took an active part in city affairs, though usually in a behind-the-scenes fashion. When the city attorney, a respected professional in

---

35. Crenson, *Un-Politics of Air Pollution.*

East Chicago's municipal government, decided to draft an air pollution law, representatives of the steel companies worked through the local Chamber of Commerce to influence that legislation. In the late 1940s a city attorney's initiative, combined with housewives' mounting but nonpolitical complaints and steel company recognition, managed to elevate dirty air from an unpleasant fact of life to a political issue demanding authoritative resolution.

Not until the mid-1950s, however, did Gary, a nearby city with similarly severe air pollution, experience the birth of the issue. Crenson credits Gary's delay to the overwhelming presence of a single steel company in that city. United States Steel not only was the city's principal employer, but it also created Gary, mapped out its roads, conferred its name, and supplied it with a major portion of its tax revenues. But U.S. Steel had a corporate policy of hands-off when it came to specific city problems. At first glance, this would appear to lower a barrier to issue creation. In actuality, U.S. Steel's detachment from Gary's problems retarded the emergence of the air pollution issue, for it was extremely difficult to get city politicians and editors to take air pollution seriously so long as U.S. Steel would not acknowledge it as a serious community matter. Power in this instance was critical because it was so concentrated and yet so detached.[36]

---

36. A decade later Gary, Indiana, had a black mayor, Richard Hatcher. Air pollution remained a condition in the city, but Hatcher's administration was handicapped in the leverage it could use on U.S. Steel. What controls were imposed on the corporation came from the federal government. When a company controls a city's job pool, "black political power" is severely limited. In January 1975, rather than comply with air pollution standards or pay a court-determined fine, U.S. Steel announced it would shut down ten of its open furnaces in Gary, resulting in the layoff of 500 workers. See Edward Greer, "Limits of Black Mayoral Reforms in Gary: Air Pollution and Corporate Power" (Paper delivered at the annual meeting of the American Political Science Association, New Orleans, September 1973); and also Greer's *Big Steel, Little Steal: Limits of Black Mayoral Reform in Gary, Indiana* (New

Thus, when seeking to determine why pollution is an important issue in Sweden but is only creeping onto the political stage in Brazil, France, or Israel, one should investigate power structures as well as cultures and experience. It may be just as important to explain why certain matters never get onto a polity's agenda as it is to explain why certain agenda items are resolved in one group's favor rather than another's. To a significant extent the politics of pollution in recent decades has been the *politics of omission,* the politics of keeping pollution off the government's agenda. Furthermore, in calculating the likely *life span* of the environmental issue in any country one must take into account the ability of power holders in and out of government to prolong the issue's life or squash it before it has barely reached adolescence. In politics as well as medicine, all deaths are not from "natural causes."

York: Monthly Review Press, 1973). Elsewhere Greer has argued that the federal government, whose intervention on the side of Mayor Hatcher could restore the local balance of power, has been unwilling to exert pressure on U.S. Steel. See Edward Greer, "Obstacles to Taming Corporate Polluters," *Environmental Affairs* 3, no. 2 (1974): 199–220.

# MOBILIZATION OF ENVIRONMENTAL INTEREST GROUPS

CHAPTER 2

## ISSUES AND GROUPS

Issues and groups are like the proverbial chicken and egg: it is likely to be a group that transforms an event into an issue; but until something is viewed as an issue, groups may ignore it or lie dormant.

There is also a reverse dynamic: abortive groups can mean the demise of an issue. Public distraction with other issues or mere boredom with environmental problems can cripple existing groups, costing them members and money. As people reinterpret their needs, groups are thrown off balance and must scramble to redefine members' common bonds. Ironically, however, if an interest group is *too* agile it may begin to blur in the public mind and dilute its impact.

A common pattern of group issue development in environmental politics is as follows: (1) an existing group without a specific concern with environmental deterioration discovers that a change is occurring that jeopardizes its interests; (2) the group is forced to stretch its old self-image and redefine its collective goals; (3) the group's new activities then alert additional organizations, some of which will turn out to be opponents; (4) the newly mobilized opponents may frustrate the environmentalists in their immediate objective, but the conflict itself will give the issue new public saliency.

Such a "scenario" unfolded during the public battle over Consolidated Edison's proposed construction of an electrically pumped storage plant at Storm King on the Hudson River. It may have been the first genuinely environmental—as versus a more traditional "conservationist"—issue in the United States and demonstrates how issue definitions and group identities shape one another.[1]

The Storm King controversy began in 1965 and remains unresolved in 1975. In 1963 New York's large utility company, Con Edison, applied to the Federal Power Commission (FPC) for a license to construct a hydroelectric plant at the foot of Storm King Mountain, forty miles up the Hudson River from New York City. If it had called for no explicit decision the matter might never have become an issue and the resultant group mobilization might never have occurred. For in any political system thousands of actions take place without an explicit authoritative choice. Without hearings, memos, or votes, these actions are "invisible" and hard to elevate into issues. Still other public decisions are made, but only by a handful of men with a narrow and specialized audience who behave as if the process were private.

---

1. The material for this case comes largely from Allan R. Talbot, *Power Along the Hudson* (New York: E. P. Dutton, 1972).

Under these conditions, too, it is difficult to create an issue. The simple fact, therefore, that Con Edison could not go ahead with its ambitious plans without a formal licensing decision by the FPC was no guarantee that Storm King would blossom into the decade's first environmental battle. What made the difference was that the Storm King project (1) would infringe on the interests of people who lived outside the immediate locale and (2) that these nonlocals commanded resources with which to capture public and bureaucratic attention.

Affluent New Yorkers with second homes on the Hudson who looked upon the area as a place for recreation pulled together an assortment of gardening and outdoor groups to form the Scenic Hudson Preservation Conference. They were not interested in the polluted waters of the river below or the relationship of energy development to air pollution; their goal was simply to preserve that dramatic bit of land whose history was intertwined with its residents back to the Revolution. The goal of the new alliance was limited: to halt the FPC licensing of Con Edison's generator. Their preenvironmentalist argument before the FPC stressed the beauty and historical significance of Storm King. Despite their impressive credentials and contacts in Washington, they failed in their initial encounter. In 1965 the FPC granted the utility its license. On the other hand, with their access to media and their skills at public communication, members of the Scenic Hudson group did arouse others' interests in Storm King's fate. It was on its way to becoming something more than a local question. Older, more broadly based organizations with an interest in what in the mid-1960s was considered conservation became involved: the Sierra Club, Audubon Society, and the Isaak Walton League.

If it had been a local garden club that had been the protestor, the issue might have remained stunted. But the resources of the Scenic Hudson group and their new allies allowed them to seek an appeal in the federal courts to overturn the FPC decision. It is this *series* of decisions, made possible when a political system has a complex network of power centers, that is crucial for sustaining an issue. In systems that are more centralized and cohesive it is harder to keep issues alive. Moreover, a group with less political sophistication or fewer lawyers among its members might simply have let the FPC's decision stand, and the Storm King dispute would have remained one over scenic beauty rather than environmental protection.

As the Con Edison license lay in limbo, mayors of the Hudson River towns, New York State officials, congressmen, and federal bureaucrats mobilized. The more diverse the contestants became, the more broadly was the issue defined. In this sense, it may take conflict to expand an issue. Some mayors worried about their river towns' economic declines and believed the Con Edison project would pump needed tax revenues into their regions. Downriver, however, New York City officials were preoccupied with increasing their own low-cost electrical energy resources. In Albany Governor Nelson Rockefeller, who had founded his career on economic growth for the state, was wooing the powerful labor unions and saw electrical energy as a necessity for expanding jobs. A second Rockefeller, Laurence, came at the issue from still another perspective but with similar consequences. Laurence was best known for his leadership of several conservationist organizations. At Storm King he saw the chance to pursue a rational "multiple use" strategy that would permit Con Edison to go ahead while still preserving the natural beauty of the site. Thus the various conservationist groups were aligned on different sides of the question, and internal debate between them was one of the things that forced stiffer questions to be asked about the real meaning of "conservation," "scenic beauty," and ultimately "public interest." Among the institutional groups that came to have a stake in the Storm King controversy were Consolidated Edison, the Army Corps of Engineers, the Atomic Energy Commission, the Department of the Interior, and the FPC. Each defined the issue according to its own peculiar bureaucratic role. Each was used to decision making out of the glare of public attention.

The Scenic Hudson group employed two strategies to stop the FPC and Con Edison; both strategies also altered the basic issue and eventually the interest perception of the group itself. First, it sought other allies, such as the largely blue-collar Hudson River Fisherman's Association. Second, it maneuvered to redefine the question so that the pro-utility FPC would no longer have the final authority.

It took the mobilization of environmental groups and the creation of a national issue consciousness to establish agencies that would perceive energy projects in wider environmental terms; until then, the contestants confronted agencies with more specialized issue outlooks. The FPC had received its mandate from the Congress under the Federal Power Act. It did not perceive the Storm King licensing decision as a matter concerning the environment. It therefore resisted giving environmental groups legal standing before the agency. In the eyes of the FPC commissioners and staff, *they* represented the public interest; the Scenic Hudson lawyers were redundant intruders. To alter the definition of the issue, then, the opponents of Con Edison had to win recognition of their own legitimacy. This could be done by bringing in other decision makers who would not evaluate the plaintiffs solely in terms of their relationship to power policy. Scenic Hudson won a ruling in federal court that confirmed the value of scenic beauty and instructed the FPC to hear public groups concerned with the "aesthetic, conservational, and recreational aspects of power development."[2] Five years later, when the United States created the Environmental Protection Agency and similar agencies were established in Europe and Asia, it would be much easier for interest groups defining issues environmentally to win standing in specific policy disputes.

The fishermen allies of Scenic Hudson broadened the issue to include ecological effects on wildlife. They also prompted the Congress's involvement, giving the contestants an additional platform on which to publicize the issue. Richard Ottinger, a Democratic congressman from a district on the Hudson, took up the cause of his fishermen constituents and initiated hearings on Storm King before the House Subcommittee on Fisheries and Wildlife Conservation, chaired by a Louisiana congressman who happened to be particularly fond of the sort of fish endangered by the Hudson project.[3] Congressional hearings were staged in the New York district. Mayors from Hudson towns who would not be directly benefiting from the project now voiced their criticisms of the unions, the chambers of commerce, the utility, and the governor, all of whom were pressuring them to approve the project. The congressional hearings also brought the Department of the Interior into the dispute, though Secretary Stewart Udall, a future spokesman for environmental control who then was anxious to avoid any confrontation with two such powerful individuals as Nelson and Laurence Rockefeller.

What started as a localized, bureaucratic, routine decision made out of public view had turned into a broadly defined public issue involving a variety of voluntary associations, economic and regulatory institutions, and levels of government. Not merely the heightened visibility, but the motleyness of the disputants served to

2. Ibid., p. 131.
3. Ibid., p. 139.

escalate and broaden the issue. The concrete result was the passage of Ottinger's 1966 Hudson River Compact Law, supported by Udall, which gave the Department of the Interior review power over all federally related projects along the Hudson. There was no guarantee that federal bureaucracy would handle ecological questions more wisely than state agencies, but the issue was taken out of its strictly New York State and energy development contexts. Bureaucrats and engineers thereafter had to be alert to the issues behind the specific questions of plant citings. Finally, in the process of refining their strategies, the protest groups moved far away from merely aesthetic or recreational definitions of their case.

As for the Storm King plant, ten years after Con Edison's original licensing request, it remains unbuilt and entangled in legal arguments. Between 1964–74 the American environmental movement grew, inflation raised the projected cost of the plant from $165 million to $500 million, and the "energy crisis" cast a new light on the matter. Despite all the money and time spent by environmental opponents, Con Edison seemed on the verge of winning in 1974.[4] Mobilizing the public around an issue, therefore, is no insurance of ultimate victory for environmentalists. But broadening existing groups' perception of what is at stake and stimulating allies as well as oppo-

---

4. *New York Times*, 6 January 1974. Scenic Hudson, the Hudson Fishermen's Association and the Sierra Club continued to spearhead the court fight, though now they had the new Federal Water Pollution Control Act to stand on. Using new legal resources that were partly the result of their own lobbying, the parties were able to get the federal courts to impose tighter restrictions on Con Edison and the FPC. In April 1974, Con Edison began construction on the 2-million-kilowatt pumped storage plant. But the following month the Federal Court of Appeals unanimously decided in favor of the Natural Resources Council and ordered an immediate reopening of the Storm King hearings on the ground that the danger of killing the Hudson's fish life had not been adequately considered. This setback for the utility came just when it was seeking a financial bail-out from the New York State Legislature. *New York Times*, 9 May 1974.

nents to participate in the policy debate may legitimize a new set of political questions and give the defeated groups easier entrée into policy circles next time.

## ENVIRONMENTAL ORGANIZATIONS AND POLITICAL CULTURE

Group formation is not as "natural" in some countries as in others. Moreover, what are deemed legitimate tactics for a group in one society may be labeled unacceptable or even criminal in others. Is it surprising, for example, that Mexico does not have energetic environmental groups despite the air pollution that now blankets its capital city? Does it take more than smog to create an environmental movement? It requires popular and governmental acceptance of public mobilization outside the bounds of regime-sponsored events. Studies of Mexican political beliefs and values conclude that Mexicans generally "do not relate easily to abstract or impersonal organizations but only to the individual who leads the movement." [5] Such feelings are not uniquely Mexican, of course. Mexicans' *personalismo* makes it difficult to start and sustain groups that lack such high-profile figures. A self-centered view of issues and the citizen's role also encourages a Mexican to underscore the *differences* between himself and his fellows, whereas an American or Canadian is more likely to stress the interests shared. In matters such as air pollution, interest-group formation depends heavily on individuals' ability to see shared needs. When, as in 1971, air pollution restrictions are imposed in Mexico, they are likely to be the result of presidential initiatives rather than environmental group lobbying, and there are likely

5. R. E. Scott, "Politics in Mexico," in *Comparative Politics Today*, ed. Gabriel A. Almond (Boston: Little, Brown, 1974), p. 256.

to be loopholes for the privileged classes who dominate the nation's politics and discourage broader participation.[6]

In France, too, interest groups are not an integral part of the political culture. This may explain in part why, despite its highly industrialized economy and its involvement in such environmentally disputed projects as the building of the Concorde supersonic transport and atmospheric nuclear testing, France lacks environmental group activity equal to that of Britain, the United States, or Japan. Studies of French politics reveal a "general aversion to associational life," coupled with a "cult of the small." [7] Thus interest groups that are formed focus on specialized interests. In addition, Frenchmen's relatively high level of ideological awareness frequently divides potential allies and undercuts common fronts on such multifaceted issues on the Concorde or atomic testing.[8] At the stage where the chief problem is to arouse public awareness of the issue, interest groups with a flare for public relations are critical. But France's political culture favors, instead, a role for interest groups which eschews

---

6. *Washington Post*, 11 July 1971. However, by May 1974 the Mexican Chambers of Commerce and the government's environmental agency both claimed their considerable compliance to the clean air law. Moreover, Mexican politicians and industrialists both seem to be increasingly sensitive to international pressure for environmental control. These were among the findings of Kirsten Haring noted in an unpublished Mexican field report, typescript, Clark University, August 1974.

7. Henry W. Ehrmann, "Politics in France," in Almond, *Comparative Politics Today*, p. 145. See also Ehrmann's *Organized Business in France* (Princeton, N.J.: Princeton University Press, 1957).

8. For instance, a continuous conflict in French history has concerned the proper role of the church in politics. Although the vast majority of Frenchmen are Catholics, anticlericalism has been a key plank in many ideological platforms. When the French bishops and cardinals came out in 1973 in opposition to France's South Pacific nuclear tests, the French naval chief of staff and several generals retorted not only with a defense of France's military policy, but also with a slap on the wrists of the clergy, warning them to stay out of public policy matters. *New York Times*, 18 July 1973.

public relations activities. Instead, most French groups are more comfortable dealing through contacts with the executive or members of the central bureaucracy. Though organizations of various persuasions contribute to electoral campaigns and parliamentary lobbying, their effectiveness is diminished by the general uncertainty about their legitimacy in the eyes of the public and the politicians.[9]

Sweden contrasts with both France and Mexico. Its political culture is remarkable for its positive evaluation on associational activities. Sweden's leading role in recognizing environmental issues and originating governmental measures is due not only to the Swedes' historic feeling for nature, but to their penchant for organization as well. One foreign observer had commented, "The propensity [of Swedes] to organize for all sorts of purposes is nothing short of incredible." [10] Swedes themselves say that their society is *genomorganiserad*, saturated with organizations. Yet there is a Swedish deference to civil servants and planners which reduces the potential impact that groups can have on specific environmental decisions. Furthermore, Swedish economic power is highly concentrated, with fifteen families and two corporations holding majority control in two

---

9. Ehrmann, in Almond, *Comparative Politics Today*, p. 147. A notable exception to this pattern was the 1974 presidential campaign of ecologist René Dumont. Dumont admitted that he had virtually no chance of winning the election, but he wanted to take advantage of the liberal television air time allowance for even minor candidates to publicize his environmental program. When out on the campaign trail Dumont told voters, "I didn't come here to influence votes . . . I came to show my solidarity with the farmers of Larzac." The farmers in this barren, rocky region were resisting plans of the National Defense Department to create a military training ground in Larzac, thus displacing 107 sheep farmers and their families. Dumont was a socialist but felt that the leading socialist candidate, Mitterand, was insufficiently sensitive to environmental problems. *New York Times*, 5 March 1974.

10. T. L. Johnston, a Scottish economist, quoted in Richard F. Tomasson, *Sweden: Prototype of Modern Society* (New York: Random House, 1970), p. 242.

hundred large Swedish industrial firms. This economic elite is not as visible in interest-group contests as, say, business firms and associations are in the United States. The result is that while organizations abound, Sweden's mode of politics encourages environmental policy making through behind-the-scenes consultations between business executives and government officials.[11]

Political cultures are modifiers, but they are not perfect insulators. An officer of one active environmentalist organization in Sweden has written, "We borrow many ideas from Friends of the Earth and Sierra Club, using their books *Environmental Handbook* and *Ecotactics*. On the local level our weapons mostly are demonstrations, signs and posters, leaflets, campaigns in the local newspapers, letters to the local authorities, teach-ins, hearings, etc." On the national level, he says, the group works on coordinating these local activities, writing letters to the concerned authorities and to the political parties, and meeting with those Members of Parliament that express particular interest in environmental issues.[12] He points also to the advantages that Swedish environmental groups enjoy over their American counterparts. "For several reasons," he writes, "I think it is more easy to raise public opinion in Sweden than in the U.S. Sweden is a small country with eight million inhabitants. We have three or four nationwide newspapers and only one radio and television company. This makes it rather easy to make your message known all over the country." Furthermore, "Since Sweden is a small country it is also more easy to get contact with the politicians—[even] for relatively small groups like our organization." [13]

Recently the Gallup Organization conducted an opin-

11. Ibid., p. 224.
12. Personal correspondence with the author, Stockholm, September 13, 1973.
13. Ibid.

ion survey among young people, aged 18–24, in eleven countries. Since youths are most likely to be the ones responsible for future environmental decisions and are in the age strata perhaps most attuned to new ideas and norms that will modify traditional political cultures, the Gallup findings are revealing. Tables 2.1 and 2.2 show wide divergences between national groups. They show differing degrees of tolerance of individual rights and freedom to organize to protest government policies. Youths also differ in their evaluations of governmental priorities, all of which affect one's propensity to join groups and to direct group activities toward politically relevant environmental decisions.

Comparative analyses suggest that certain sociocultural conditions may be especially important in making interest groups a critical factor in environmental politics.

1. The society is homogeneous enough to generate mutual trust among citizens, but pluralistic enough to prompt individuals to organize for the promotion of particular interests.
2. Industrialization and urbanization have made the problems of environmental control visible and the resources—literacy, skills, numbers, money, media—for organization readily available.
3. Government is perceived as effective enough to protect the environment and resolve conflicts among citizens.
4. There is a widely held belief that life can be bettered through collective action.
5. The current regime tolerates groups not created or coopted by it. The average citizen is confident of that toleration.

*Table 2.1.* **Multinational Survey on Attitudes of Youth, Survey of Youth 18–24 Years, Summer 1973.**
**A. General**

*Gallup Opinion Index*   October 1973

*Question:* Often people who are forced to leave their land and homes to make way for public works projects organize protest movements. Which of the following describes your feeling toward such people most accurately?

| | Brazil | France | India | Japan | Phil. I. | Sweden | Switz. | U.K. | U.S. | W. Ger. | Yugo. |
|---|---|---|---|---|---|---|---|---|---|---|---|
| These movements often are simply the result of selfishness | 19% | 12% | 49% | 29% | 18% | 13% | 17% | 20% | 21% | 28% | 73% |
| These movements are a justifiable defense of human rights | 80 | 81 | 50 | 69 | 80 | 80 | 82 | 75 | 77 | 62 | 27 |
| No answer/No opinion | 1 | 7 | 1 | 2 | 2 | 7 | 1 | 5 | 2 | 10 | — |

*Source:* This data is drawn from "Multinational Survey on Attitudes of Youth," *Gallup Opinion Index*, October 1973, p. 25.

*Table 2.2.* **Attitudes of Youth: B. Government**

*Gallup Opinion Index* Summer 1973, October 1973

*Question:* I am going to read some statements about our national government or society. For each one, please tell me whether you think it is "true" or "false."

| | Brazil | France | India | Japan | Phil. I. | Sweden | Switz. | U.K. | U.S. | W. Ger. |
|---|---|---|---|---|---|---|---|---|---|---|
| The government is placing too much emphasis on the benefits of the nation as a whole at the cost of individuals | | | | | | | | | | |
| True | 55% | 68% | 71% | 88% | 55% | 68% | 63% | 68% | 74% | 44% |
| False | 44 | 13 | 27 | 11 | 42 | 27 | 35 | 28 | 25 | 49 |
| Na/No[a] | 1 | 19 | 2 | 1 | 3 | 5 | 2 | 4 | 1 | 7 |
| The government's strong emphasis on industrial development tends to make people unhappy | | | | | | | | | | |
| True | 30 | 42 | 39 | 90 | 43 | 76 | 64 | 65 | 69 | 47 |
| False | 69 | 39 | 58 | 8 | 54 | 20 | 35 | 32 | 29 | 46 |
| Na/No[a] | 1 | 19 | 3 | 2 | 3 | 4 | 1 | 3 | 2 | 7 |

| | Brazil | France | India | Japan | Phil. I. | Sweden | Switz. | U.K. | U.S. | W. Ger. |
|---|---|---|---|---|---|---|---|---|---|---|
| The government sometimes goes in the opposite directions from those in which the people really want it to go | | | | | | | | | | |
| True | 71% | 76% | 75% | 85% | 62% | 86% | 66% | 90% | 87% | 56% |
| False | 27 | 9 | 24 | 13 | 34 | 10 | 32 | 8 | 12 | 38 |
| Na/No[a] | 2 | 15 | 1 | 2 | 4 | 4 | 2 | 2 | 1 | 6 |
| Present society places heavier emphasis on rules and laws than on confidence among men | | | | | | | | | | |
| True | 78 | 78 | 83 | 75 | 71 | 83 | 82 | 79 | 81 | 69 |
| False | 20 | 11 | 15 | 24 | 26 | 13 | 16 | 18 | 17 | 25 |
| Na/No[a] | 2 | 11 | 2 | 1 | 3 | 4 | 2 | 3 | 2 | 6 |
| In the present grossly materialistic society, money reigns supreme | | | | | | | | | | |
| True | 77 | 90 | 86 | 84 | 75 | 82 | 90 | 86 | 88 | 78 |
| False | 21 | 5 | 13 | 16 | 24 | 15 | 10 | 12 | 11 | 16 |
| Na/No[a] | 2 | 5 | 1 | — | 1 | 3 | — | 2 | 1 | 6 |

[a] Na/No represents No answer/No opinion.
*Source:* See table 2.1.

## TYPES OF INTEREST GROUPS

Just as we have divided countries between those having political cultures supportive of interest groups and those in which cultural norms dampen group activity, so we can distinguish types of groups according to how they will handle political issues. The three sets of categories are: (1) specific versus general (or multiissue) groups; (2) extragovernmental versus intragovernmental groups; and (3) objective versus subjective groups.

*Specific versus general groups.* Its members, as contrasted with those in more diffuse groups, are concerned about environmental policy or at least some corner of the debate because the group's raison d'être was clearly spelled out when they joined. A narrow organization may run into trouble if it tries to move into seemingly tangential arenas. Its rank and file may lag behind or resign. There was a split within the U.S. Sierra Club, for instance, over just this question: How broadly should the club define its interests? Should it stick to its orginal dedication to parks, conservation, and recreation or should it become involved in broader issues of urban environment and energy development? Likewise in the Soviet Union there may be disputes among the intellectuals in the loosely organized "civil rights movement" over whether to concentrate on such conventional libertarian issues as literary censorship or take up the cause of ethnic minorities or polluted Lake Baikal as well. Sacrifice of specificity may be worth the price if it is rewarded with wider membership, greater media attention, and diverse political entrées. The complexity of the environmental issue makes it rare for concerned groups to keep their initial tight focus. An oil company, for example, at one time may have been interested only in the relationship of petroleum exploration to the environment, but later will invest in fertilizer and pesticide production and thus become a powerful voice in any environmental debate of agricultural policy.

*Extragovernmental versus intragovernmental groups.* The second set of categories divides groups between those existing outside of government and those with official status within government. The lines can be hard to distinguish. Governments spin off an endless series of quasi-official consultant groups or commissions, composed of nongovernment personnel but possessing official access, staffs, and budgets. This is one way for systems that stop short of fully nationalized industry—Sweden, United States, Britain, Japan—to exert governmental control over economic planning. Though often it is questionable as to who is controlling who. Such advisory groups multiply in the environmental field because of its technical complexity. How they should be categorized depends more on the members' own perception of their role than on formal definitions.

Official bodies have self-interests—the essential one being institutional survival. Governmental agencies can behave like voluntary interest groups, though being on the inside bestows advantages. Institutional interest groups are more assured of *access* and *information,* two key resources in any political competition. Environmental debates entail alliances between insiders and outsiders— e.g., between the Ministry of Mines and the coal companies or between a fledging Environmental Agency and conservation groups—so that this categorization does not predict on which side of an environmental issue a group will fall. Rather, it predicts what sorts of resources a given group will bring to the contest. When voluntary associations are curtailed, as in many authoritarian systems, governmental agencies become the principal actors in environmental policy making.

*Objective versus subjective groups.* An objective

group is one that exists "in the eyes of the beholder." For instance, ski-resort developers in Italy represent a group insofar as an observer lumps them together because of their common enterprise. Likewise, "meat inspectors," "town planners," and "strip-mining executives" are objective groups. You may be able to predict how each will react to certain legislative proposals because of what you know about the needs of individual ski-resort developers or the needs of coal company executives. But an individual objectively labeled a "ski-resort developer" by an outsider may not think of himself as a member of an organization which will mount a campaign to kill the legislation. By contrast, a subjective group's members are conscious of belonging to a collective body and have lines of communication and explicitly shared goals allowing them to act in unison. As an environmental issue gains saliency, subjective group awareness will grow, often dividing people. In Japan, for instance, eaters of fish and chemical industrialists both became subjectively group-conscious when the alarming news of mercury poisoning of fish made headlines. Business firms with the most to lose in the short run from tight pollution controls had a head start in formally organizing, while victims of pollution may never be more than objective groups. Though Japanese taxi drivers, housewives, and civil servants share a common dietary dependence on fish, they may be slower to transform this objective fact into actual cooperation.

## STRATEGIES AND RESOURCES

Two sets of determinants shape a group's approach to a particular environmental issue: the group's own resources and the openness of the political system. Group resources include (1) its treasury, (2) the numbers and spread of its membership, (3) the degree of unity among its members, (4) its access to information, and (5) the seriousness with which it is regarded by the media and by policy makers. The nature of environmental issues is such that ability to draw upon technically trained experts can prove critical. How many biologists a group can recruit may be more important in an environmental debate than how many voters it has on its rolls. Where policy making occurs mostly within bureaucratic agencies, groups with the best-informed and most prestigious experts have a distinct edge. Environmental politics, as well as the politics of weaponry, make it clear that scientists are not apolitical. Their research grants, their professional status, and often their jobs depend on political decisions. Thus individual scientists and professional scientific associations are becoming familiar political actors in countries as disparate as the United States, the Soviet Union, and Brazil.[14]

Groups need substantial funds if they must rely on advertising to awaken the public to the dangers of pollution. Unfortunately, it takes money to raise money, and many environmental groups never accumulate enough to mount an effective drive to ask for more. Money is also crucial if the group relies on court suits to alter business or bureaucratic policies. Lawyers, researchers, and transcripts are costly, especially when the case may run on for several years, as did the Storm King controversy.

---

14. For analysis of scientists in politics, see Marlan Blissett, *Politics in Science* (Boston: Little, Brown, 1972); Daniel Greenberg, *The Politics of Pure Science* (New York: New American Library, 1967); H. Gordon Skilling and Franklyn Griffiths, *Interest Groups in Soviet Politics* (Princeton, N.J.: Princeton University Press, 1971); Kalman H. Silvert, *The Social Reality of Scientific Myth* (New York: American University Field Staff, 1969). The latter analyzes the role of scientists in Third World nations.

Numbers of members will be valuable if the organization is trying to revise environmental policies by changing the compositions of a city council or a national legislature, whose delegates run in relatively small districts suitable for voter mobilization. Numbers can be vital as well if environmental issues are placed on the ballot, though referenda victories call for alliances among several groups and cooperation of nonmembers in order to produce a majority.

The second set of determinants refers less to the attributes of the interest group than to the characteristics of the policy-making system. Where are the crucial environmental decisions being made? Is decision making on these matters so fragmented that a group must win access to a variety of ministries or local councils? How open are the governmental institutions to outside petitioning—e.g., are there formal hearings in most agencies? To what extent do elected officials run on issue platforms rather than on the basis of personal charisma or factional ties? A multinational group such as Friends of the Earth finds that resources relevant in Britain may have minimal impact in France. One FOE correspondent noted that the organization's penchant for symbolic demonstrations, which caught media attention in London, were abortive in Paris. In fact, one FOE "bike-in" in France was not even tolerated by the local police as a legitimate form of petitioning.[15]

"Lobbying" refers to the exertion of group pressure on policy makers. But the location of the "lobby" may be outside a legislative committee room, instead of outside the general legislative chamber. Or the proverbial lobby could be removed from the legislature altogether, located in the halls of executive departments. With the demise of legislatures in many countries, interest groups may find

15. Colin Blye, "Friends of the Earth," *Ecologist* 3, no. 2 (February 1973): 71.

that lobbying is effective only if it is directed at bureau-crats. But bureaucrats are less open to petitioning groups; their careers do not depend as directly on public support. Thus, only groups that can prove their official legitimacy and professional expertise are likely to get inside these bureaucratic "lobbies."

Environmental groups have a stake in how politicians envisage career ladders. They have a stake, too, in reforms that alter those career ladders. Reforms proposed for the U.S. House of Representatives, for instance, could upset the folk wisdom that committee specialization is the road to a successful political career—and the best target for special-interest lobbying. Critics argue that the U.S. congressional committee system has made certain legislative areas not simply accessible to, but captives of, special interests and senior congressmen. Tight relation-ships between a committee's members, bureau adminis-trators, and certain interests prevent competing groups or the general public from receiving a fair hearing.

Reforms such as those proposed by the Bolling committee in 1974 would reorganize U.S. House commit-tees topically—e.g., jurisdictions for national parks and forests would be taken from the House Interior Commit-tee and given to a new Agriculture and Natural Resources Committee, while water and air pollution would be transferred from the Public Works Committee to the Interior Committee. But would this reform alone solve the more fundamental problem of congressional-lobbyist co-ziness? To open up the congressional committee process to newly mobilized environmental groups there must be reforms that require congressmen to rotate from commit-tee. Rotation would ensure that the Interior Committee would not become the instrument of western state repre-sentatives while northern urban members shunned envi-ronmentally related committees in favor of the Education and Labor or the Banking and Currency Committees.[16] Lacking such reforms, groups shut out from one branch of government may adopt strategies to penetrate other sectors of government. American environmental groups have been compelled to do this when unable to exert decisive influence on regulatory agencies such as the FPC. They turned their attention to the courts. Decades earlier, similarly frustrated Negro civil rights organizations pursued the same strategy. In fact, the Environmental Defense Fund deliberately emulated the NAACP's successful Defense Fund. Among the most effective environmental groups employing litigation to accomplish what they could not do through the legislature and regulatory agencies have been Ralph Nader's several public-interest groups. One of their principal weapons has been the National Environmental Policy Act of 1969 which required agencies to submit environmental impact statements before launching projects.[17] The Nader groups have also made full use of the Freedom of Information Act to get courts to compel agencies to open their files to interested citizens.

When in England, however, Nader found that audiences viewed skeptically his advice to use the courts, since English courts are not as active policy makers as they are in the United States. In Japan the courts have been important for victims of industrial pollution because of recent legislation making polluters liable for compen-

16. Norman J. Ornstein, "Committees: Case for Rotation," *Washington Post*, 6 January 1974. The House of Representatives eventually did defeat one wide-ranging Bolling reform package, passing instead a compromise bill that would have less impact on existing committee jurisdictions. *Washington Post*, 9 October 1974.
17. See Robert W. Burchell and George Hagevik, *The Environmental Impact Handbook* (New Brunswick, N.J.: Center for Urban Policy Research, Rutgers University, 1974). See also: Walter A. Rosenbaum and Paul E. Roberts, "The Year of the Spoiled Park: Comments on the Court's Emergence as an Environmental Defender," *Law and Society Review* 7, no. 1 (Fall 1972): 33–60.

sating persons who can prove injury. But where such laws are not on the books or where the courts deliberately refrain from poltical involvement, groups will find such alternative strategies less productive. They may have to concentrate on the legislature at least to the point of securing laws that then give environmentalists a standing in court.

To judge fairly the effectiveness of a given environmental group, one clearly must take account of these structural as well as cultural limitations. For what appears to be a feeble effort for an American organization may be a significant accomplishment for a group facing (1) centralized and closed bureaucracy, (2) an ineffectual legislature, (3) timid courts, and (4) business opponents enjoying quasi-official status in the government.

Thus at first glance it appears that French environmentalists put up a shamefully weak opposition to the supersonic transport compared to the successful effort mounted in the United States. But the anti-SST Americans had more independent media outlets (French television is government-managed), more politicized scientists, a more potent legislature, and more money. Furthermore, though American environmentalists confronted powerful aerospace lobbyists having close connections with government, their French counterparts were in competition with an aerospace industry *owned* by the government. Jean-Jacques Servan-Schreiber's newsweekly, *L'Express*, one of the most outspoken vehicles for anti-Concorde opinion, bemoaned the "regrettable" absence of debate among French political parties, parliamentarians, and government's Economic Council. It noted that, whereas in both Britain and the United States there was a lively public discussion of the financial and social costs of the SST, in France there was no acknowledgment of the potential conflict between contemporary necessities and the illusory imperatives of "progress," no explicit governmental priorities, aside from the dream of restoring France's "greatness."[18] In both the United States and Britain, cities whose employment depended on SST construction acted as powerful pressure groups behind the SST (Filton in Britain, Seattle in the United States).[19] But in France, the city of Toulouse, with a similar business and labor stake in the government project, had far more potent pro-SST allies and few effectively organized opponents.[20]

## CLEAVAGES AND COALITIONS

A group's true interest may not be revealed in its title or mandate. "Oil" companies have large investments in coal and natural gas; clubs for the benefit of auto drivers can be dependent on the favor of auto producers instead.[21] Genuine interest definition grows out of (1) what the group relies on for its survival and (2) the sorts of information group leaders receive with which to determine their optimum position on a given issue. Interest definition also derives from the adaptability of the group.

Gross categories such as "farmer," "youth," or "labor" are of minimal use in predicting actual group alignments on a particular policy question. First, these

18. "Le Prix de Concorde," *L'Express*, 10–16 May 1971, pp. 60–64.
19. For a description of Concorde's impact on Filton, see *New York Times*, 3 February 1973.
20. The contrast between Britain and France is repeated in their respective public reactions to above-ground nuclear testing. In Britain the antibomb movement was equivalent to the civil rights movement in the United States in terms of popular levels of mobilization. While the largest Paris turnout against the French bomb test in the spring of 1973 was a meager 1,000 participants—"including a heavy lacing of foreigners, mostly Britons." *Washington Post*, 22 July 1973.
21. For charges that the American Automobile Association abandoned its auto driver members' interests and colluded with auto makers in environmental controversies, see *Washington Post*, 10 June 1973; Bert Schwarzchild, "Auto Club Revolt," *Sierra Club Bulletin*, June 1973, p. 18.

broad objective groupings can have different economic and political implications in different nations. Second, within each category are usually important subdivisions that affect the rewards or losses that a policy choice can imply. Thus, "farmers" in the United States may refer increasingly to large landowners or corporate agribusinesses, which require huge applications of petrochemical fertilizers for profitable production. In Thailand, "farmers" may refer in the main to small landowners or tenants who have neither the acreage or the capital to commit themselves to such chemically oriented agriculture. To hope or expect, therefore, that an international coalition of farmers would unite for or against United Nations agriculture policies would invite disappointment.

Similarly, "businessmen" are not monolithic. Because business firms are the principal polluters in most countries (e.g., industrial pollution accounts for one-third of all solid waste, one-half of all air pollution and more than half of total water pollution in the United States), they are usually cast as opponents of stiffer environmental laws.[22] But the business world in both capitalist and socialist economies is complex. Greek shipping magnates with enormous investments in oil-carrying supertankers and Italian businessmen who have invested in the Riviera tourist industry both are anxious to protect investments and maximize profits. But this shared characteristic is likely to put them on opposing sides at international conferences called to impose restrictions that will reduce oceanic oil slicks. Businessmen are not by mere definition opponents to all antipollution legislation. Some businessmen are making profits off the new regulations, by contracting with government agencies or other firms to produce newly required environmental technology that reduces industrial pollution. Such businesses can be very useful "bedfellows" for environmental lobbies.

---

22. Environmental Protection Agency estimates quoted in *New York Times*, 30 June 1974.

Some of the deepest cleavages among youth and other social groups in environmental debates occur along economic-class lines. Environmental organizations in most countries are heavily middle class. Working-class interest in and support of antipollution policy has been minimal. A principal exception may be in Japan, where mercury poisoning, smog, and garbage disposal all have mobilized blue-collar workers, fishermen, and farmers. In the United States perhaps the most notable proponents of environmental protection outside the middle class have been the American Indians. Among the few American working-class organizations fighting pollution are Chicago's Citizens Against Pollution (CAP) and the Chicago-Calumet Environmental Health Committee (CEOHC). It has been when environmental issues have been tied to *urban* discontents and to *job* conditions that the environmental movement has overcome its bourgeois parochialism.[23]

Politics makes strange bedfellows. The environmental movement inspires motley coalitions because environmental interest groups in most countries have too little money, too few governmental connections, and too few members to operate alone. This inadequacy is exacerbated by the superiority of resources possessed by the environmentalists' usual opponents on matters of regulation. In addition, environmental issues themselves encompass a wide assortment of policy questions and thereby provoke loose, ad hoc coalitions. Many of them are not dependable outside of a small range of policy matters. Beyond them, the coalitions crumble.

23. See Staughton Lynd, "Blue Collar Organizing: A Report on CEOHC," *Working Papers* 1, no. 1 (Spring 1973): 28–34; Edward Greer and Paul Booth, "Pollution and Community Organizing in Two Cities," *Social Policy* 14, no. 1 (July–August 1973): 42–49. For a description of how U.S. environmental law can be used by urban poor to advance their own cause, see Daniel J. Kramer, "Protecting the Urban Environment from the Federal Government," *Urban Affairs Quarterly* 9, no. 3 (March 1973): 359–68.

Elections may most necessitate coalition building, but they produce the least reliable results. The problem comes on the morning after. For beyond getting the environmentally sensitive candidate elected or ratifying the environmentally beneficial proposition, the coalition has to put in a solid enough showing at the polls to make the mandate clear to policy makers. "Reading the mandate" is what really counts in politics, and a coalition may fail to make a strong enough impression to ensure that the politicians will read the environmental "message" in the election returns.

The Norwegian election on the Common Market is a case in point. A loose coalition of environmentalists took an active part in the September 1972 election, arguing that by joining the Common Market Norway would only be taking another step toward unlimited industrialization. In the end, Norway, unlike neighboring Denmark and Ireland, voted not to join the Common Market. But when 54 percent of the electorate rejected entry it was no assurance of a success for the Norwegian environmentalists. For other groups had waged vigorous campaigns against Common Market entry as well. When it came to reading the mandate, trying to decipher the "message" that the voters were sending to their policy makers, the environmental message was difficult to separate and thus difficult to act upon. The People's Movement Against the Common Market, organized by Arne Haugestad, called "Norway's Ralph Nader," brought together Norwegians who had never allied before—farmers and city workers, pious Lutherans, student radicals, conservative nationalists, and environmentalists. They faced EEC proponents in the form of the two major political parties, trade-union leaders, large and small businessmen, the state civil service, and the state television. The very heterogeneity of the coalition which gave it victory in the voting booth made it hard to hold together afterward. Then each group wanted their own interests to be read as the major determinant in persuading voters.[24]

Postelectoral politics raise other problems for successful coalitions. If the target of the combined efforts has been a specific candidate or party slate of candidates, the postelection problem becomes one of monitoring the sponsored official's future decisions and holding him accountable. This sort of interest-group activity requires more political commitment than activism in a short-run campaign. It tests the durability of an electoral coalition, for agreement among diverse groups over a candidate's general platform can disintegrate when concrete policy decisions are made by the winner in office.

## CONCLUSION

The interplay between groups and issues in environmental politics has meant that neither has stood still. Accurately analyzing environmental interest groups means keeping track of a dynamic situation. The simple categories of specific-general, intra- and extragovernmental, and subjective-objective should help explain why groups with similar environmental stances behave differently in the political process. The categories should also help in tracing the transformations of a single group as it broadens or narrows its focus, as it is coopted by government or is divorced from officialdom. Each transformation will alter its stockpile of resources and thus its potential effectiveness.

In this age, when McDonald's hamburgers thrive in Tokyo and Korean ski boots warm feet in Vermont, it is all too easy to imagine that group strategies are instantly interchangeable. If a fishermen's blockade works in Ja-

24. *New York Times*, 1 October 1972; *Boston Globe*, 1 October 1972.

pan's polluted harbors, why not try it in San Diego? If suing government bureaucrats elicits policy changes in the United States, shouldn't it be a winning technique in Britain? Indeed, information, resources, and tactics have been exchanged across national boundaries and have helped internationalize the environmental movement. There is nothing automatic in such political trade, however. The ability of environmentalists (and their antiregulation opponents, who also exchange tactics and information) to assimilate foreign organizational techniques will depend on the political culture—especially public willingness to join nongovernmental groups and government elites' tolerance of such activism. It will depend also on the complexity of the exporter's and importer's political systems—especially diversity of access points into official policymaking.

Finally, certain distinctive characteristics of environmental issues affect the form and resources for related interest groups. First, environmental debates usually are technical. Groups lacking their own technical experts are handicapped. Second, environmental issues are not easily contained; they slip over into other issue areas. The result is that environmental groups are constantly seeking allies even though the political marriages may be short-lived. Third, environmental issues are all too frequently cast in terms of a dichotomy between aesthetics and leisure on the one hand versus growth and jobs on the other, with the consequence that in most countries environmental groups have suffered from a middle-class bias. They often find themselves facing a formidable alliance between business and blue-collar labor. In such a match environmentalists rarely have the resources to alter government policy effectively.

# BUREAUCRATIC POLITICS IN POLLUTION CONTROL

## CHAPTER 3

## THE BUREAUCRATIC IRONY

Bureaucracy existed long before the modern state, in the Chinese, Ottoman, and Russian empires. But the common attributes of bureaucracies are peculiarly modern: hierarchical relations, salaried employees, standardized rules of procedure, impersonal authority, specialization, and dependence on written files for continuity. Environmental movements in many countries contend that modern life has alienated man. Technology alienates man from nature, while bureaucracy alienates man from man. Bureaucratization and pollution seem to be extensions of the same phenomenon: modernization.

Ironically, once a society reaches a level of environmental deterioration that necessitates control and plan-

**76**

ning, it is bureaucracy which takes the responsibility for implementing those controls and drafting those plans. Environmental movements may reduce pollution, but rarely bureaucracy. What environmentalists can do, nevertheless, is make bureaucracy more accountable, less preoccupied with its own institutional interests. But this is not reducing bureaucracy; it is reform coupled with expansion.

Where environmental conditions have achieved issue status, new bureaucratic agencies have been created: Poland, Malaysia, Brazil, Yugoslavia, the United States, and Britain, to name but a few. Sometimes a new agency is a token gesture and offers no assurance that the issue will be tackled. Elsewhere, the new agency is subsumed under an existing ministry where it experiences uneven competition for funds and authority. In a small number of cases the environmental agency is blessed with autonomy, budget, and political support to implement current regulations effectively and to take the lead in proposing new legislative mandates for itself. These cases to date are few. That is, a new bureaucratic agency can itself be the victim of the pitfalls of bureaucratization.

Sweden represents a country that has experienced the ironic choice between the two sorts of modern alienation. In the early 1970s Swedes moved to curb environmental abuse and at the same time tried to dismantle their pervasive, centralized, and elitist bureaucracy. One scholar estimates that a set of bureaucrats numbering "no more than one-hundred" situated in eleven ministries are responsible for most policy making.[1] An ombudsman to handle public complaints against bureaucrats and the requirement that official documents

---

1. Martin Schiff, "Welfare State Bureaucracy and Democratic Control in Sweden and Its Implications for the United States" (Paper presented at the Northeastern Political Science Association, Buckhill, Pa., November 1973), p. 2.

be publicized are two mechanisms that curtail potential and administrative arbitrariness.

The Rikstag (parliament) elections of September 1973 suggested that the traditionally accepted bureaucratic system was losing some of its legitimacy in Sweden. What makes the 1973 election all the more interesting to a student of environmental politics is that the second major issue in the campaign was the destruction of the Swedish landscape. The Center party (formerly the Farmers party), led by a sheep farmer turned politician, ran on a platform of nostalgia for the preindustrial countryside and yearning for the nonbureaucratic life. Party leader Thorbjorn Falldin exuded the spirit of a simpler Swedish past when the streams were cleaner and relationships less highly organized.[2] The Social Democratic incumbent, Prime Minister Olof Palme, summed up the mood behind the election: "There is a deep-seated dissatisfaction about industrial life. It creates problems—movements of populations, new towns, a shift from the countryside and so forth. And the Center Party stands for a mood, a nostalgia for the days when people didn't have to move to the cities and work in factories. In a sense, Falldin somewhat represents the past."[3] In reality, Falldin and his party took up specific issues dominating the Swedish *present,* tougher environmental regulation and bureaucratic decentralization.

The election, in which 90 percent of the electorate voted, left the Social Democratic party in power, but barely.[4] The Center party won 25.1 percent of the votes. Morning-after commentators noted that, though the Swedes are essentially a conservative people who resist change, they were warning the Social Democrats to take

2. *New York Times,* 19 September 1973.
3. Schiff, "Welfare State Bureaucracy," p. 13.
4. Ibid.

their environmental and bureaucratic discontents seriously and to respond with genuine reforms.[5]

This interweaving of environmental concerns with hostility toward bureaucracy has afflicted some environmental advocates with a political blindness. They ignore the intricate workings of bureaucracy that affect environmental policies, preferring to focus on the politics of legislatures and elections. The most effective environmental activists in many countries are those who carefully analyze bureaucratic politics and provide support or prodding to potential allies within bureaucratic departments.

Inaccessibility and unaccountability are age-old bureaucratic sins. Bureaucracies thrived because they supplied professional expertise and institutional continuity that was not affected by the uncertainties of popular politics. These attributes, nonetheless, also have been shortcomings in the view of citizens who expect government to be not merely competent and stable, but responsible as well. Although environmental agencies have been created largely in response to public demand, there is no guarantee that, once instituted, those new departments will not fall prey to the isolating reflexes that mar so many older organizations. Environmental activists thus have two seemingly contradictory tasks to perform with regard to the bureaucracy. They must give moral support (and sometimes technical assistance) to infant environmental agencies struggling for funds and intragovernmental cooperation. At the same time, activists must constantly monitor those agencies to ensure that they remain committed to environmental protection and accessible to the general public.

5. Ibid. See also "Sweden: Which Way Now?," special supplement in the *Economist*, 13 April 1974.

## INSTITUTIONALIZATION OF
## ENVIRONMENTAL AGENCIES

"Institutionalization" refers to the process through which an organization gains legitimacy and becomes agile and resourceful enough to survive changes of regime or policy. It is a process, too, by which an organization achieves a routineness so that persons within it relate to one another in regular patterns, not according to capricious personal likes or dislikes. Most environmental agencies are less than a decade old and not yet fully institutionalized. Some may never achieve institutionalization. Some fail to win confidence and cannot cope with political change; their mandates remain too vague and budgets too marginal for them to develop regular procedures. Eventually such agencies may be dismantled or simply shunted aside and denied any meaningful role in policy formation or implementation.

A new environmental agency faces the problem of rivalry with older bodies that possess authority in areas of environmental importance. Such established departments—e.g., departments of forestry, mines, agriculture, commerce—are more firmly institutionalized and have the resources and political contacts to ward off intrusions from any bureaucratic upstart. Because of the diffuseness of environmental problems a new regulatory agency is likely to be cast in the role of a governmental "poacher," a role that undermines its efforts to gain legitimacy and cooperation. Environmental policies are "generally incompatible not only with many of the conventional beliefs and behavior patterns of most people in industrialized countries but also with the institutional arrangements of government and the economy." [6] A new bureaucratic

---

6. Lynton K. Caldwell, "Environmental Policy and Public Administration," *Policy Studies Journal* 1, no. 4 (Summer 1973): 211.

organization will be superimposed on preexisting arrangements, usually with the aim of modifying the routine orientations of those arrangements. For instance, the environmental department may be charged with persuading the Ministry of Industrial Development to think less of raising the national GNP or persuading the Ministry of Agriculture to measure their success by criteria other than the output of wheat or rice or cotton. The environmental agency usually has a multidimensional task that does not jibe neatly with the existing unidimensional, mission-oriented organizations.[7]

In addition, the new environmental administrator has a more nebulous constituency than most of his bureaucratic colleagues. He has "all the citizens" as his constituents, since environmental disruption threatens the well-being of everyone. But an experienced bureaucrat would prefer a less encompassing, better organized, and more easily discernible interest sector in his court. "All the citizens" are hard to mobilize, have diffuse interests, and short attention spans. Those environmental organizations that support and monitor newly created environmental agencies are giving the agency a more concrete and bureaucratically "useable" constituency. An effective agency is one that is not "underprivileged" due to lack of specific clientele interest and yet is not so much the captive of one client that it is robbed of its integrity.

### Origins and Efficacy

An environmental agency can be held responsible for its actions, but it cannot be blamed (or rewarded) for the circumstances of its creation. The political culture and existing bureaucratic milieu into which it is born may handicap it from the outset. For instance, Italy's bureaucracy has been characterized as inefficient, baffling, and

---

7. Ibid.

nightmarish. The national bureaucracy includes some one million people, most of whom are recruited by letters of recommendation from politicians or from among citizens afflicted by natural or man-made disasters. After a six-month trial period they do not have to work unless they want to. Nobody on the state payroll can be fired unless jailed for a criminal offense. Given this situation, plus the instability of the party regimes in Italy, one should not be surprised that the Italian government "lost" the several-million-dollar grant donated to it by international organizations for the protection of Venice, a city endangered by water subversion and polluted air. Italians, in fact, have a category of state bookkeeping known as *residue passivi:* "money allocated for public investment but never spent because all trace is hopelessly lost in some ministerial bog." An estimated $15 billion had been lost in this fashion according to 1973 figures.[8]

One feature of an agency's origin that may prove important, especially when it treads on the territory of preexisting departments, is its political sponsorship. The French Ministry of the Environment was created by a 1971 decree from President Pompidou; Pompidou likewise selected a fellow Gaullist and energetic director of his party to head the new body.[9] Similarly, in Poland the Communist party leader and prime minister, Edward Gierek, took a personal role in a blue-ribbon commission

---

8. Claire Sterling, "Italy," *Atlantic*, October 1973, p. 16. See also William Murray, "Letter from Rome," *New Yorker*, 8 April 1974; *New York Times*, 4 June 1974. In an analysis of the bureaucratic causes of the 1973 cholera outbreak in Naples, William Murray found that the Ministry of Health was not even informed by local officials of the epidemic but learned of it first from the newspapers. William Murray, "Letter from Naples," *New Yorker*, 26 August 1974, p. 55.

9. "How France Attacks Its Environmental Problems," *World*, 13 February 1973, p. 41. Pompidou's successor, Giscard d'Estaing, however, selected for his environmental minister a man not known to be close to him. Giscard also raised questions about the future efficacy of the ministry when he renamed it the Ministry for the Quality of Life, a title that made its mission less clear. *New York Times*, 29 May 1974.

on the environment. Gierek's own power base is in Silesia, one of the most industrialized and polluted regions of Poland.[10] Since 1971, Poland, rated one of the most polluted countries in Europe, has allocated large budgetary sums to the ministry that grew out of that blue-ribbon commission, the Ministry of Regional Planning and Environmental Control. It began its institutional life with a $775 million budget for the 1971–75 period covered by the government's five-year plan.[11] By 1973, reports showed that air pollution had dropped in Poland by 4 percent compared with 1970 levels, and water pollution had remained almost constant despite a 10 percent industrial growth. Still, scientists agree that one-third of Poland's water is unusable and another third is substantially polluted.[12] Thus the new ministry will have to continue to enjoy considerable political party support if it is to make headway in containing and reducing the nation's pollution.

The person appointed to serve as the first director of a new agency can have a lasting impact on it. The most visible international bureaucracy dealing with environmental policy is the United Nations Environment Program (UNEP). The term "program" was adopted deliberately because there were doubts about its being bureaucratized within the regular UN Secretariat. The hope was that designating it as simply a program would keep it open and flexible.

---

10. Dan Morgan, "An Engels Vision Blurred by East Europe's Pollution," *Washington Post*, 21 February 1971.
11. *New York Times*, 17 September 1972. See also "Environment," *Polish Perspectives* (Warsaw) 16, no. 9 (September 1973): 63–64.
12. "Poland Giving High Priority to Environment," *Environment News* (EPA, New England region), January 1973, p. 2. However, in another field, coal mining, Poland now has a distinct advantage environmentally. Polish hard coal has a low sulphur content which makes it superior to most American coal. During his 1974 visit to the United States Prime Minister Gierek discussed American purchases of Polish coal. *New York Times*, 6 October 1974.

The agency had the advantage of having as its first director the man who worked directly with the 130 separate nations that were to be serviced and advised by UNEP, Maurice Strong, a Canadian. Strong utilized that personal prestige in an early conflict with the parent bureaucracy, the UN Secretariat, headed by Secretary-General Kurt Waldheim. Waldheim had submitted a sharply reduced budget for the fledgling UNEP without even showing it to Strong. Not only did Strong disagree with some of the cuts, he also apparently suspected that the budgetary maneuvers reflected the secretary-general's desire to keep control over the new agency in the New York headquarters. Strong, on the other hand, had insisted on a certain freedom from bureaucratic control from the center and, perhaps for this reason, had not been as opposed as many Westerners to the decision that UNEP be situated in Nairobi, Kenya, far from New York.[13] Strong was in better position than most heads of new agencies to shape the grounds for its beginning. He used his personal prestige and the public identification of UNEP with himself and threatened to resign if the budgetary and control questions were not resolved more to fit his own conception of UNEP's role. The secretary-general backed down, presumably under the knowledge that Strong, "one of the most popular and effective figures in the history of the United Nations," had the votes to win his point were it to come to a General Assembly ballot. Still, UNEP was created in such a way that even Maurice Strong as first director would have "more responsibility than authority" and would be free to select only part of UNEP's senior staff.[14]

13. "At the UN—A Showdown and Two Important Decisions," *World*, 30 January 1973, p. 69. See also *New York Times*, 7 December 1972.

14. Ibid. For a report on Strong's frustrations with bureaucratization, see Jeff Radford, "Planet Savers in Nairobi," *Nation*, 21 December 1974, pp. 655–56.

Resignation by a prominent bureaucrat may have to be not only threatened but also carried out in order to sustain the new bureaucracy. In West Germany in 1972, Dr. Bernard Grzimek resigned as head of the federal government's three-year-old wildlife protection commission. He told Prime Minister Willy Brandt in his letter of resignation that he objected to the light fashion in which ecology issues were treated in the prime minister's programmatic message, to the low budgets allotted for his office, and to the lack of cooperation from other federal ministries. The commissioner also complained that West Germany was "far behind" Britain, Sweden, and the Netherlands in protecting the environment. He was reported to have believed that his very prominence in the West German public's mind, through his popular television programs, was being used by the government to give the illusion that it was doing more than in fact it was.[15] Popularity notwithstanding, Commissioner Grzimek's resignation was accepted by the regime. Yet, prodded by his departure and his call for a sixteen-fold increase in expenditures on environmental protection, the government unceremoniously announced an increase in the budget for nature preserves and wildlife protection from $300,000 in 1969 to $3 million in 1973. This may have been as much a retort to the Grzimek charges as a genuine remedy to the conditions behind them. German scientists have voiced growing concern about the extinction of dozens of animals and plants once native to Germany.[16]

## BUREAUCRATIC MANDATES

The existing political culture, the character of the

---

15. *New York Times*, 11 February 1973.
16. Ibid.

initial support from politicians, the leverage and independent stature of the first administrator, and the adequacy of the first budget—all will affect an environmental
agency's future effectiveness. In addition, and perhaps
most fundamental, is the legal *mandate* that defines its
mission and authority. Three different sorts of bureaucratic weaknesses can derive from an original mandate.
First, the new body may be given too broad a mission
with insufficient resources. It is likely to lose its credibility and be dismissed as a bureaucratic "paper tiger." The
American Environmental Protection Agency (EPA)
suffered such damage when it was strapped into a legal
straitjacket by the 1970 Clean Air Act of 1970 requiring
the EPA to ensure that all states achieve healthy air by
1975, with possible extensions to 1977. In response to this
ambitious mandate, the EPA administrators proposed
stringent parking and automobile-use controls on the
nation's most air polluted urban centers. It then faced
angry public hearings at the local level and eventually
was compelled to modify drastically or to surrender many
of its original proposals. The experience did nothing to
enhance the political leverage of the EPA. Likewise, on
lacking the necessary influence to compel the auto manufacturers to produce the emission-exhaust controls that
would allow the 1975 deadlines to be met, the EPA had to
return to Congress to ask for an extension of its own
deadlines.[17]

Second, a defective mandate is one that defines the
agency's mission too narrowly for it to have a meaningful
impact. Frequently coupled with the narrow mission are
very modest staffs and authority. The Soviet Union, for
instance, in 1973 announced the creation of a high-level
agency to monitor air and water pollution nationwide. But
the new bureau was given no powers to resolve contradic-

17. "The EPA's Contribution to Better Cities," *Business Week*, 4
August 1973, p. 21.

tory directives from other parts of the bureaucracy and/or the conflicting demands for growth as well as environmental protection that come from the Soviet leadership. Forest management operations in the Volga basin demonstrated the impotence of the new monitoring service. Scientists in the Ministry of Forestry have the authority to determine how much timber can be cut without endangering the erosion-preventing function of the forest. The timber-cutting quota is assigned, but with an eye to the demands of the Ministry of the Lumber and Wood Processing Industry, which is more concerned bureaucratically about lumber production than soil erosion and agricultural conservation. Thus the Ministry of Forestry becomes more an exploiter than a protector as was originally intended. The new service has no legal machinery to untangle these conflicts.[18]

Finally, a mandate may be so ambiguous as to paralyze the new agency. There are occasions when an ambiguous mandate in the hands of a politically talented and bureaucratically energetic administrator can produce a more wide-ranging program than the bill's authors imagined. But more typically, such ambiguity makes a bureau vulnerable to outside criticism and plagues it with internal caution. Observers have found a mixture of both reactions to ambiguity in the American political and bureaucratic response to the National Environmental Policy Act of 1969. Open to varying interpretations has been the act's requirement, to be supervised by the EPA, for submission of environmental impact statements. The result has been numerous court suits. One federal judge wrote that the act's meaning is "more uncertain" than that of most statutes because of the generality of its phrasing (*Hanley* v. *Mitchell*, at 20217). Another justice presiding over such a court battle, commented that the

---

18. "How Much Did You Say?," *World*, 10 April 1973, p. 36.

act is "so broad, yet opaque, that it will take even longer than usual to fully comprehend its import" (*City of New York* v. *U.S.*, at 20276).[19] The very vagueness, however, permitted the interest groups monitoring the EPA's bureaucratic performance to push the agency further. Friends of the Earth and others charged that the EPA's Grants Administration Division had given out more than $5.3 billion between 1970 and 1973 for the construction of sewage treatment facilities and yet had required no environmental impact statements from its own grantees. The interest groups sought out the author of the EPA's mandate, Congressman John Dingell of Michigan, who informed the EPA that he interpreted the NEPA as requiring impact statements for EPA projects as well as from those of industries and other bureaucratic departments.[20]

Original sponsorship and mandate do not predetermine all future administrative actions. But the circumstances surrounding and shaping the creation of an agency do leave their mark. Before blaming a department for organizational timidity or special-interest bias, one needs to examine closely the political dynamics out of which the agency evolved. Altering bureaucratic behavior may call for amending the legislative mandate or undoing knots of interests that designed a weak agency in the first place.

## "WHERE YOU STAND DEPENDS ON WHERE YOU SIT"

To threaten resignation, as did UNEP and West

19. Leslie L. Ross and Richard A. Lixroff, "Political Dynamics and Environmental Policy," *Policy Studies Journal* 1, no. 4 (Summer 1973): 226–27.
20. *Washington Post*, 18 November 1973.

German environmental administrators, takes conviction and courage. It also requires autonomy. What Maurice Strong and Bernard Grzimek shared were careers and reputations outside of their administrative posts. Each perceived himself as not merely an official serving under a particular regime, but as an independent spokesman who offered his services to a regime. Their publics agreed, which is why their threats of resignation carried such weight.

What we are talking about here is the concept of "role." A role is a pattern of behavior that others expect of any person in a certain position, whether that position is prime minister, civil servant, forest ranger, wife, or soccer pro. Role is not just what you think is right or natural to do in a position, but what others have come to expect of you in such a position. To change a role is to change not personnel, but popular expectations. Some roles are well-defined: the role player knows what is expected of him and what is risked by violating those expectations. Other roles are quite vague or controversial: expectations about what is needed or legitimate in a post are in flux or are contrary to one another. Should the environmental administrator, for example, be an activist lobbying in parliament for tougher controls or should he be content to implement only what is already on the books? Should he be openly partisan, a supporter of the current party in power, for the sake of increasing his leverage in top policy circles; or should he be a nonpartisan technocrat with freedom to slap the wrists of any politician that advocates environmentally harmful policies?

Pressure to adapt one's behavior to the expectations residing in a role is what is described by the formula "Where you stand depends on where you sit." [21] To a

---

21. This phrase is used by Graham Allison in his study of the bureaucratic dynamics behind the Kennedy administration's decisions

cynic the formula suggests that most officials and politicians are hypocritical and will adopt whatever policy stance makes their institutional life most comfortable. A more subtle observer understands that the same person actually will see an issue differently if he or she is in one institution rather than in another. Each institution affords a distinctive vantage point on a question. A department's particular mandate, its sources of information, its collective memories—all shape an administrator's perception. Each department over the years also has developed a set of a priori assumptions about what is significant, what is delayable, what is risky, what is proper or improper. These are transmitted to any incoming bureaucrat.

For instance, how should one explain the apparent changes of priority in Peter Walker over the last few years? Walker was an eminently successful British businessman appointed the first "overlord" of Britain's superagency, the Department of the Environment. Later Walker left DOE and was appointed head of the Department of Trade and Industry at a time when his country faced its severest economic crisis since the war. Is hypocrisy a sufficient explanation for Peter Walker's unbusinessmanlike advocacy of environmental controls on British industry? Is hypocrisy or opportunism adequate to explain his greater concern for mining high-sulphur coal while Secretary of State for Trade and Industry than with clean air during the coal miners strike of 1974? [22] Without performing an in-depth psychological study on Peter Walker, one cannot answer those questions definitively, but we do know from other studies that a person is affected in his behavior by "where he sits," that is, by what role he holds in an organizational

during the Cuban missile crisis. See Allison's *Essence of Decision: Explaining the Cuban Missile Crisis* (Boston: Little, Brown, 1971), p. 176.

22. For a more thorough analysis of Walker's performance in the Department of the Environment, see chapter 8.

environment, and that role condition may alter his sense of priorities.

Such alterations may not be dramatic. Some persons have firm personal ideas about public goals and private styles of action that move with them as they progress from one role to another. This may or may not enhance career success for the individual. For instance, Walter Hickel came to Washington as Richard Nixon's secretary of the interior. As a land developer and then as governor of the growth-conscious state of Alaska, Hickel evidenced little concern for environmental hazards. But his rigorous senatorial hearings that preceded his cabinet appointment made it clear that the old expectations of secretaries of interior were giving at least somewhat to new role expectations that included directives that the secretary be a protector of the environment as well as a booster of western states' mineral exploitation. In this regard, Walter Hickel-the-Alaskan-governor changed when he became Walter Hickel-the-secretary-of-interior. What did not change, however, was Hickel's political style of outspokenness. This suited the role of a frontier state governor, but was less appreciated by the President and White House staff when displayed in a cabinet officer who, presumably, was a subordinate to the Chief Executive. A clash of roles as much as any factor led to the resignation of Walter Hickel from the Nixon cabinet.[23]

Role perceptions take a strong hand in shaping officials' notion of their own self-interests. Frequently, a bureaucrat's role is defined implicitly as protecting the

---

23. For insight into the personal background of another American bureaucrat, a man who, as head of the Bureau of Reclamation, was a bête noire of environmentalists, see John McPhee, *Encounter with the Archdruid* (New York: Ballantine Books, 1971), pp. 135–215. Floyd Dominy, former head of Reclamation, grew up in the drought-ridden West, saw families suffer and cattle die. To him, damning the Grand Canyon was literally beautiful. There was no tension at all between his personal values and his bureacratic role.

organization against any outside interference, even from institutions formally mandated to check or monitor the agency. At other times, role is defined even more narrowly so that one's official or organizational interest is to promote one's own personal self-interest. The high tolerance for corruption in Italian bureaucracies, for instance, may be due to a conception of role which draws little distinction between self-interest and organizational interest. A scholar examining the planning process which has produced so many of Rome's severe environmental problems has observed that transferring planning responsibilities from private operators to public officials did little to increase the level of responsible planning: "Public operators, no less than private operators, were concerned to protect and enhance their particular interests and displayed a uniform lack of concern for the city's planning problems." [24] The agencies themselves were seemingly motivated by similar interest, each city and national bureaucracy seeking to maximize its own resources, especially the property values of its own land holdings.[25] This is scarcely the organizational atmosphere which encourages role definitions that emphasize public responsiveness or interagency cooperation.

Role perceptions can be changed by several means. Most require not only an awareness of the complexity of role dynamics, but also sufficient political influence to affect the changes. In some instances, expanding the constituency to which the bureaucrats are sensitive may alter the actors' role perceptions. Since individuals gain their ideas of what is proper or expected behavior from those with whom they have the most salient contact, role definitions could be altered by opening up the given

24. Robert C. Fried, *Planning the Eternal City: Roman Politics and Planning since World War II* (New Haven: Yale University Press, 1973), p. 233.
    25. Ibid., p. 234.

agency to a wider assortment of public and governmental "audiences" who have different expectations about what the officials in that agency should be doing. Changing the reporting system might compel an agency to take account of a new ministry's expectations. Transferring an organization from one department to another can have the effect of altering role perceptions. A good deal of environmental politics in many countries has involved reshuffling organizational charts for the sake of refocusing officials' attention and reordering their priorities.

Another strategy for altering role definitions is to bring a new public sector to bear on the agency. This has been a favorite tactic of Ralph Nader in his campaigns to make American bureaucracies more responsive to their environmental constituencies. In Italy, perhaps the most effective environmental interest group working in this way is the large and prestigious Italia Nostra. In 1974 Italia Nostra published a report, "The Mistakes of Rome," detailing the acuteness of Rome's environmental problems. For example, the report noted that in 1974 Rome received less than half the amount of fresh water it received through its aqueducts at the time of Emperor Augustus. Moreover, there was found to be more typhoid fever in Rome alone than in the entire United States. Rome's rate of carbon monoxide pollution, the report continued, was often six times above the health hazard limit, and the noise level in Rome was higher than any other place in Italy.[26] In a political system in which the bureaucratic agencies have been unresponsive to generalized public interests, Italia Nostra's most significant accomplishment may have been to put bureaucrats on the

---

26. *New York Times*, 3 February 1974. "Mistakes of Rome" listed among the city's inadequacies: need for three times as many sewers as now functioning; reduction of the 50,000 Romans who live in shacks or caves; expansion of park space from the current average of a little more than 20 square feet per inhabitant; need actually to enforce the now frequently violated 1962 zoning regulations.

alert that their behavior would be monitored by an interested public group which had role expectations contrary to those of the usual politician or private contractor, the city planner's or engineer's traditional audience.[27]

## BUREAUCRATIC SCHIZOPHRENIA

Bureaucracies are as rife with conflicts of interest as any assortment of lobbyists or parties. Role conflict in this context occurs when a single agency is saddled with competing or incompatible goals. Some conflicts are the price paid by the creators of the agency for coordination and consolidation of administration: a variety of ill-fitting responsibilities are housed under one roof for the sake of bringing together all administrators working on a single policy area such as health or law enforcement. In socialist political systems such contradictions are frequently built into the very fabric of the governmental structure. Bureaucratic regulators are the same persons assigned to meet or even surpass production quotas. One Soviet commentator has described this situation as "asking the goat to mind the cabbage patch." [28]

The American system has fewer instances of a government agency being simultaneously both industrial producer and regulator. But environmentalists have pointed to the Atomic Energy Commission as an example of just such role conflict. The AEC is authorized to be the nation's booster for nuclear power at the same time as it is depended upon to be the prime safeguard against hazardous utilization of nuclear power. Issue changes can

---

27. For descriptions of locally mobilized citizens groups taking corrupt Italian officials and developers to court, see *New York Times*, 27 January 1974.

28. Quoted from *Pravda*, in "World Environment Newsletter," *World*, 8 May 1973, p. 37.

exacerbate role conflict such as that afflicting the AEC. Though it was burdened with a "split personality" from the start, emergence of the environmental movement and prophesies of a national energy crisis stretched each side of the AEC almost to the breaking point.

Since role conflict undermines an agency's credibility, a director will be at pains to deny that any ambiguity exists: all the agency's responsibilities are compatible. When interviewed by an environmentalist journal, therefore, AEC director Dr. Dixie Lee Ray denied that the AEC's promotion of nuclear technology compromised its role as regulator of nuclear power. According to Ray, the AEC only develops the technology; it is then up to the electrical power utilities to use that technology in ways that meet AEC safety regulations. There was no inherent partnership between the AEC and private companies— there was no goat in the cabbage patch. Ray continued: "I think that we get caught in a philosophical bind by using words like 'promote' and 'partnership' and that kind of thing. Now the industry has to come to AEC for permits and for rules under which the power plants have to be built and operated . . . there is always the possibility when you have a regulator and a group that is regulated that those who issue the licenses might try to get cozy." But, Ray goes on to quote her predecessor, James Schlesinger, "We don't want any [licenses] fixed. Just remember this agency will work with you, but we have to provide the licenses, we have to provide the supervision, the monitoring. And we can fine." As Commissioner Ray understood her predecessor, this was "not a statement that said 'we've been in bed with you before and now we're getting out.' " [29]

Despite public assurances, Ray did recognize the tensions within the AEC. Internally, Dixie Lee Ray, who

---

29. Charles Conconi, "An Interview with Dixie Lee Ray," *Environmental Quality*, July 1973, p. 21.

was appointed in part because of her background as a marine biologist who would not be beholden to the nuclear energy business, moved to reorganize the AEC to get the goat out of the cabbage patch. She divided regulatory roles from promotion roles.[30] Externally, there were efforts in Congress to reduce AEC role ambiguity by limiting the agency's field of authority. A bill passed the House of Representatives in 1973 which would make the AEC a simple research and development agency, while creating a separate nuclear energy council to handle safety regulations.[31] If passed, this would change bureaucratic and political behavior patterns built up over twenty-seven years.

Transforming the AEC official's role perception would mean transforming others' role perceptions as well, since "role" is a set of mutually agreed upon expectations. It is not just in the mind of the AEC official, but in the minds of most persons who deal with that official. Consequently Ralph Nader, in his criticism of the AEC's alleged laxness in upholding nuclear safety regulations, pointed an accusing finger as well at the primary group from which AEC officials over the decades have received their impressions as to what their bureaucritic roles properly were: the Congressional Joint Committee on Atomic Energy. Nader's criticisms of the Joint Committee carried a theme similar to those charges launched at the AEC itself—the organization which was intended

---

30. In 1974 the AEC's regulatory arm had a staff of 390, of whom 242 were assigned to inspect the 74 reactors then operating or under construction. Another 44 staff members inspected the 8,839 license holders using radioactive materials, while 74 staff members were assigned to safeguard nuclear fuels from theft or diversion. *New York Times*, 25 August 1974.

31. *New York Times*, 12 January 1974. For a more detailed analysis of the politics of the AEC, see H. Peter Metzger, *The Atomic Establishment* (New York: Simon & Schuster, 1972). An extensive case study of the AEC's actions in promoting a nuclear reactor on the Susquehanna River is found in: McKinley C. Olsen, "The Hot River Valley," *The Nation* 219, no. 3 (3 August 1974): 69–85.

to be the watchdog had ended up being a partner instead.[32]

One has only to read the daily press to see instances of this institutional self-interest at work in other countries. In Canada, too, the policy area of energy development has generated bureaucratic role contradictions which have either hobbled ministries or made one part of their original mandate inoperative. The role contradictions are particularly severe for the Canadian Department of Internal Affairs and Northern Development. Its personnel's role is dictated by the wide variety of policy objectives it is supposed to pursue for the federal government in Ottawa. On the one hand, the department's objective is to provide a rising standard of living and quality of life for the largely indigenous residents of the harsh northern territories. On the other hand, the department sees its role as one of encouraging viable growth in that region so that the north will contribute more to Canadian national economic progress. Finally, it is mandated to preserve the north's environment and to protect Canadian national sovereignty in that region.[33] One Canadian bureaucratic observer has inquired, "How can the Department hope to promote mineral, oil, and exploration on the one hand and at the same time fulfill obligations as trustee of the northern native peoples or protector of the northern environment?" The writer provides his own reply by noting that individuals as well as bureaucratic organizations can try to resolve role conflicts by assigning priorities, in this instance, by giving prime attention to fostering economic development and neglecting its two other functions.

The problem with this resolution is that the bureaucrat's assignment of role priorities may not agree with his

---

32. *New York Times*, 29 January 1974.
33. Robert Page, "The Image of the North," *Canadian Forum*, June–July 1973, p. 8.

various clienteles' assignment of priorities. The Canadian critic offers an alternative solution, one that sounds strikingly like that offered for the AEC. "This internal Department dilemma can only be resolved by splitting up the 'empire' to end the current conflict of interest. This challenge is particularly pressing as some senior Departmental officials move back and forth between industry and government. The northern development function," according to the writer, "should be shifted to the Department of the Environment (which covers the rest of the country), and a separate Native Affairs Department should be established." He warns that "this would not eliminate the chance of further sellouts for native rights or ecological conditions but at least they would be fought out at the cabinet level between ministers, not decided privately and internally within a department by civil servants not accountable to the electorate." [34] What the AEC and Canadian Internal Affairs Department reformers suggest, however, is that role conflict can be resolved by increasing the fragmentation of governmental authority and fostering more healthy conflict within the bureaucracy.

There is another solution, though it may require more profound political change than is involved in manipulating organizational charts: resolve the conflicts of public priorities which create role conflict. This might require public choices that would involve ideological and cultural alterations rather than mere bureaucratic reform.

## BUREAUCRATIC COMPETITION

"Where you stand depends on where you sit" presumes that bureaucrats working in different agencies will

34. Ibid., pp. 8–9.

perceive issues and goals differently. How that intra-bureaucratic conflict is handled and how wide or narrow are the acceptable perimeters of any debate will vary from political system to political system. But few if any bureaucracies are not shaped by these internal differences.

Since environmental agencies are latecomers, they are frequently seen by preexisting agencies as threats or competitors. A new environmental ministry or department is a new competitor for budgetary allotments and for presidential or cabinet attention. If the agency is launched at a time of tightened governmental expenditures, then the budgetary dimension of bureaucratic conflict is apt to be especially important. If, on the other hand, the agency makes its debut in an era of expanding governmental revenues and public expenditures, then the conflict with preexisting bureaucracies may be muted. When economists or policy analysts talk about a "constant pie," they are referring to a relatively inelastic financial pool from which more numerous and insistent constituencies must be "fed."

One change in environmental politics from the 1960s to the 1970s in most North American and Western European nations is that the movement and the agencies it spawned made their appearance when government coffers were relatively generous, expanded as they were by growing industrial economies; but they met their first real political tests of effectiveness in the early 1970s when the booms were pierced and governments began to cut expenditures. When the budgetary pie is fixed, the bureaucratic actors with the most political experience and most influential allies are the best equipped to sustain their annual budgets.

To measure the "clout" a given environmental agency has, however, one must examine more than simply the "bottom line," that is, the total figure allotted to the agency. For some agencies by their very character do not require large sums to carry out their mandates. What is more indicative of an environmental agency's budgetary acumen relative to competing departments may be (1) the rate of growth of its funds from year to year and (2) the public evaluation with the adequacy of its budget as expressed by the agency heads. In these terms, the Polish environmental agency, for instance, has fared far better in bureaucratic conflict than has its counterpart in West Germany. Likewise, although the Environmental Protection Agency has become one of the major dispersers of public works funds in the American political system, its budgetary efficacy has to be questioned when it suffers presidential impoundments of large amounts of funds allocated to it by Congress.

Resources available to an agency for protecting and advancing its own bureaucratic interests will vary. Money will be a resource to use and to compete for in virtually all systems. Other resources include support from well-organized and well-connected public interest groups—a resource less important in those countries in which such groups either are suppressed or are scarce due to political culture and poverty. Also, alliances with other powerful bureaucratic agencies are a resource for bureaucratic competitors. This may be particularly important in precisely those systems where nongovernmental groups are ineffective. Personal support from a major political leader, from legislative factions or committees, or from local authorities can be used to forward an agency's interests. And finally, the saliency of that agency's field of expertise to the nation's immediate and long-range legitimized interests is crucial if the organization is to get a serious hearing in cabinet councils or budgetary sessions.

Much of the conflict among bureaucratic organizations occurs between different levels of government. One thinks immediately of the problems created by the federalist structure in the United States, Canada, and elsewhere. Studies of Switzerland and West Germany, both federal systems, reveal that a principal challenge in implementing effective environmental controls is coordinating administrative activities among the various states and between the center and the states.[35]

But federal systems are not alone in putting a high premium on center-periphery bureaucratic coordination. Most unitary states that oversee complex socioeconomic conditions such as those in urban and industrial countries also must find means to ensure that local or municipal air pollution or health officials follow standard practices. For local responsibility is growing even in unitary states as governments seek to take the burdens off the already overtaxed center and distribute more duties to levels of government closer to the people and to their immediate environments. Italy and France have introduced decentralization reforms within the last five years, and Britain is considering wide-ranging regional "devolutions" of power. Usually, however, such federalizing or decentralizing schemes increase the number and necessity of bureaucratic forms of organization. Thus, as the various cantonal governments of Switzerland began passing laws regulating air quality in their own regions, there was the necessity to create a Federal Clean Air Commission to coordinate and standardize the laws. In Switzerland this suggestion met with considerable resistance because of the historically cherished autonomy of the cantons.

---

35. U.S. Department of Health, Education, and Welfare, *Profile Study of Air Pollution Control Activities in Foreign Countries: First Year Report* (Research Triangle Park, North Carolina, 1970), pp. 53, 148–49.

## BUREAUCRATIC CLIENTELES

Interagency conflict is caused in part by the fact that officials in various departments see themselves as servants of special sectors of the nation. Sometimes these public sectors are simply in the minds of the individual bureaucrats, who perceive the "farmers of France" or the "sportsfishermen of America" or the "shipbuilders of Scotland" as their particular responsibility. In this sense most bureaucrats are encouraged to think of themselves as representatives even though they never campaign for office. Agency identification with a certain group is the product of historic evolution—what group's needs prompted the government to create the agency in the first place. Identity also results from recruiting practices in which an agency relies on one sector of the public for its personnel because that sector has the training and interest that attracts them to the agency. Identity can be shaped as well by experiences in political conflicts during which one sector of the public has consistently provided support for the agency when it was under attack from other quarters. It is hard for a dedicated career civil servant not to look with sympathy upon a group that has come to the agency's defense when its budget was about to be slashed or when it was about to be merged with some larger department by an unhappy parliament.

The dynamic of clientele-bureaucracy relationships is commonly described as "pressure." Just as often, in fact, the dynamic is one of mutual support and reinforcement. Portraying all interactions between administrators and interest-group spokesmen as adversary relations misses part of the reality of bureaucratic politics.

When the relationship becomes especially reinforcing and has a routineness that allows one to distinguish it from merely random or occasional interaction, we can refer to the relationship as a "complex." This is the meaning of the term "military-industrial complex" first coined by President Eisenhower. It refers to a recurring pattern of mutually rewarding relationships between certain government agencies and their primary nongovernmental clienteles. A "complex" often will include elected politicians as well. In the American context the notion of a complex—or "subgovernments" as they are also called—violates the Madisonian assumption that politics in a free society is generated by conflict between groups vying for government's attention and benefits. A complex, on the other hand, implies behind-closed-doors collusion. Moreover, a complex contradicts the constitutionally designed system of tension between legislature and executive and put in its stead selected subsystems of mutual support between interest legislators and specialized civil servants, with the public and even the President often excluded. Finally, and most relevant to this discussion, the idea of a complex presumes not only that bureaucrats have interests of their own, but also that they have the means to mobilize public support to further those interests.

Outside the United States, the idea of a complex or subgovernment may not be as heretical, depending on how much close interaction between government and private sectors is presumed legitimate and natural. The Swedish, British, and Japanese, for instance, believe that consensus gained through careful consultation and minimal conflict or coercion produces more efficient governance. In socialist systems a complex is almost the definition of what managerial-political relationships are intended to be, except if they begin to subvert the party's program.

Every time a chemical company executive phones a bureaucrat, one is not witnessing a complex or subgovernment at work. A complex has to be shown to exist

with careful empirical findings that demonstrate *regular,
distinctive, mutually supportive,* and *exclusivist* relationships between specialized governmental and extragovernmental sectors. It takes time for a political interaction to
blossom into a genuine complex. Analysts have begun
to compare the American, Japanese, and Soviet systems
to determine whether "military-industrial complexes"
can be said to exist in any or all.[36]

In the environmental area several investigators have
asserted that bureaucrat-politician interest-group complexes do exist in the United States. Frequently, however,
these arguments are based on coincidental or projective
evidence rather than on demonstration of actual joint
decision making. This may reveal political dealings that
fall short of the public ideal and yet not merit the label of
"complex." [37] One study which comes closer to substan-
tiating the existence of a complex focuses on the implementation of U.S. government regulations for the safe
levels of asbestos pollution in factories. In a long and
thorough investigation journalist Paul Brodeur found that
indeed Department of Labor industrial safety inspectors,
allegedly independent scientific testing laboratories, and
chemical companies such as Monsanto worked in concert
to keep legal definitions of health hazards lower than the
actual health requirements for workers justified. Brodeur
labels this mutually rewarding relationship which compromises effective antipollution control in American fac-

36. See, for example, Fritz Vilmar, "The Military-Industrial Com-
plex in West Germany and Consequences for Peace Policy," *Journal for
Peace Research,* no. 3 (1973): 251–58; William T. Lee, "The Politico-
Military Industrial Complex of the USSR," *Journal of International
Affairs* 26, no. 1 (1972): 73–86; Vernon V. Asparturian, "Soviet Military-
Industrial Complex—Does It Exist?", *Journal of International Affairs*
26, no. 1 (1972): 1–28.

37. Among the best of the "muckrakers" revealing such suspicious
coincidences of association and agreement among polluters and regulators is the American author James Ridgeway: *The Politics of Ecology*
(New York: E. P. Dutton, 1971); *The Last Play* (New York: New
American Library, 1974).

tories the "medical-industrial-government complex." [38]
Though the "complex" is an extreme form of bureau-
clientele relations, it reflects the more routine effort of
administrators to cultivate sympathizers. In a study of the
ineffective U.S. National Air Pollution Control Administration (NAPCA), Charles Jones calls Francis Rourke
"everlastingly correct" when the latter says:

> A first and fundamental source of power for adminis-
> trative agencies in American society is their ability to
> attract outside support. Strength in a constituency is
> no less an asset for an American administrator than it
> is for a politician, and some agencies have succeeded
> in building outside support as formidable as that of
> any political organization. The lack of such support
> severely circumscribes the ability of an agency to
> achieve its goals, and may even threaten its survival
> as an organization. [39]

Jones goes on to describe in detail how the NAPCA
failed to muster that outside support and consequently
was ineffectual as an implementer of antipollution regulations. It required the escalation of pollution as a general
political issue in the United States, together with the
mobilization of all groups specifically interested in clean

38. Paul Brodeur's detailed study of asbestos regulation appears in
his *Expendable Americans* (New York: Viking, 1974). It also was a
five-part series in the *New Yorker* under the title "Annals of Industry:
Casualties of the Workplace": 29 October 1973, pp. 44–106; 5 November
1973, pp. 92–142; 12 November 1973, pp. 131–77; 19 November 1973, pp.
87–150; 26 November 1973, pp. 126–80.

39. Charles O. Jones, "The Limits of Public Support: Air Pollution
Agency Development," *Public Administration Review* 32, no. 5 (Sep-
tember/October 1972): 502–8. In a later article Jones contends that the
Environmental Protection Agency and its supporters in Congress have
been able to go beyond cautious incremental policy steps to undertake
"great leaps" in preventive regulation because the EPA, unlike its
predecessor, has extensive public support. Charles O. Jones, "Specula-
tive Augmentation in Federal Air Pollution Policy-making," *Journal of
Politics* 36, no. 2 (May 1974): 438–64. See also Charles O. Jones, *Clean
Air: The Policies and Politics of Pollution Control* (Pittsburgh: University of Pittsburgh Press, 1975).

air, to revive the now faltering NAPCA.[40] On the other hand, an investigatory team sponsored by Ralph Nader concluded that the same NAPCA *did* have outside clients, but they happened not to be committed to clean air as much as they were to minimizing governmental controls. Nader's team concluded that it was the NAPCA's unwillingness to confront or offend industrial corporations that was the reason for the air pollution agency's timidity.[41] So much of environmental bureaucratic activity takes an adversary form. Agencies impose standards, threaten fines, conduct monitoring inspections, require technology changes, reject proposed projects. It is true, of course, that the environmental agency usually also has some carrots to accompany the sticks, but positive inducements usually are quickly forgotten when an industrial firm fears its smokestacks will be closed or its river dumping publically reprimanded. Bureaucrats, like most people, would prefer to have daily dealings which are relatively tension-free. It makes their job easier if they are not constantly viewed with suspicion or anxiety or resentment. This natural desire to minimize the unpleasantness in job relationships is increased when the bureaucrat believes that his department's effectiveness depends on voluntary acceptance. Many agencies are created with only meager sanctions and with procedures so potentially crammed with red tape that compliance imposed through sheer coercion would be either ineffective or grossly time-consuming. One of the realities of bureaucratic regulation of the environment is that the persons that the official must deal with day after day are the persons he must regulate. The potential offender, then, becomes the key to whether the bureaucrat's life is smooth or harried,

40. Ibid.
41. John C. Esposito, *Vanishing Air: Nader Report on Air Pollution* (New York: Grossman Publishers, 1970).

and he is likely to adopt behavior that promote smoothness. Following this inclination, environmental officials in West Germany, for instance, make a point of consulting the very industrialists they are supposed to be inspecting.[42] Likewise, regulators in Sweden actively promote a spirit of cooperation and mutual confidence between their agencies and industry.[43]

But where is that fine line that separates government-industry cooperation for the sake of more effective regulation from government-industry coziness that makes the bureaucrats life less tense but leads to environmental neglect? The line is rarely spelled out in any laws. To determine whether cooperation serves the environment better than conflict, one probably has to look at specific decisions to determine whether the public interest has been sacrificed. There also probably should be some third-party participation to limit the insulated coziness that can easily evolve over time between regulator and regulated. This third party may be a legislative oversight committee that is independent of bureaucracy and industry. Or it may take the form of "public" members on the regulatory commission, persons who have the technical expertise to evaluate any compromises raised, but whose self-interests are not enmeshed in the bureaucracy's institutional interests or in an industry's profits. Two additional "third parties" are a superior administrative overseer, which monitors the monitors, and an energetic press which seeks to open any conveniently closed doors to smoke-filled rooms.

"Smoke-filled rooms" sounds as though they are places for selfish deals, where money is passed under the table or corruption hangs in the air. But they may be the

---

42. "World Environment Newsletter" *World,* 22 May 1973, p. 41.
43. U.S. Department of HEW, *Profile Study,* p. 130. See also Joseph Board, "Legal Culture and the Environmental Protection Issue: The Swedish Experience," *Albany Law Review* 37, no. 4 (1973): 603–31.

places where bureaucrats and industrialists try to reach accommodations that merely reduce each other's work load while compromising the public's health. In Canada, the National Energy Board has been accused of being "industry's best friend." The charge was not of illicit bribery, but of complacency. Instead of playing an adversary role and developing information of its own, the NEB commissioner consistently underrated Canada's energy requirements because they didn't bother to challenge the energy industry's assurances that fuels exported to the United States would not leave Canadian consumers short. The NEB saved themselves work and concern in the short run, but in the long run such complacency meant that more of Canada's terrain had to be exploited to meet the unanticipated demands.[44]

### CONCLUSION

Is the image of the harassed bureaucrat simply a public myth? Are all the alleged "pressure groups" that seek appointments with government officials not there to pressure but to support or protect or accommodate? While it is true that a great deal of bureaucratic politics is motored by these activities, pressure and conflict also exist. A truer picture of environmental politics, especially when viewed comparatively, is of some agencies better able than others to insulate themselves from pressure by building up satisfying relations with their prime clienteles and by warding off intrusions from the prying eyes of politicians and other public groups. On the other hand, there are agencies that have faithful supporters but still face public exposure and political accountability. These agencies will lead lives mixed with accommodation and

---

44. Ian McDougall, "The NEB: Let's Give Credit Where Credit Is Due," *Canadian Forum*, June–July 1973, pp. 30–31.

pressure. The latter will escalate when the agency is responsible for laws that are particularly controversial and which are at the center of an emerging issue. As the issue is defused and the aroused public is diverted to other concerns, the agency will find itself able to go back to its more comfortable relationships.

Thus the extent to which any agency's bureaucratic dynamic is characterized by clientele pressure or accommodation will depend on several factors: (1) the visibility of the agency due to a current issue debate; (2) the effectiveness of institutional mechanisms in supplying "third parties" to bureaucratic negotiations; (3) the administrative pride and sense of assurance which permits an agency to take the risk of angering a regular client; (4) the existence of a variety of interest groups, not just one or two, which are interested enough in the agency's field of action to keep track of its decisions, appear at hearings, challenge its procedures.

Even a cursory investigation of bureaucratic politics surrounding environmental policy reveals the prevalence of conflict within the bureaucracy and between bureaucrats and outsiders. Such conflict appears to frustrate efficient environmental planning and regulation; it encourages permanent ad-hocism and the sacrifice of policy in the name of strategy. A list drawn from these assorted illustrations would include among conditions that *minimize* bureaucratic conflict:

1. No personally ambitious politicians are appointed heads of bureaucratic agencies.
2. An explicit and uniform ideological commitment is the criterion for all career civil servants.
3. All bureaucrats, regardless of their special technical expertise or professional allegiance, come out of a common educational and social-class background.

4. The political party controlling the cabinet and legislature permeates the institutional structure as well.
5. The legislative, judicial, or executive bodies that monitor the bureaucratic agencies are relatively united and homogeneous.
6. Organizational techniques exist which are sophisticated enough to provide for adequate means to ensure interagency communication and coordination.
7. The society and economy regulated by the various agencies are not marked by serious cleavages.
8. The revenues taken in by government grow as rapidly as the expenditure demands so that most agencies feel confident of receiving the budgetary allotments they need to sustain their own programs.

Quite clearly, even in this bureaucratic utopia, environmental agencies can be 90-pound weaklings. For harmonizing conditions could serve to downplay the government's responsibility for treating or preventing pollution. Sometimes, then, the very existence of the conditions for bureaucratic conflict give a new environmental agency a fighting chance, though they may simultaneously breed delay, waste, and distrust in governmental operations.

# UNDERDEVELOPMENT AND THE "LUXURY" OF POLLUTION

CHAPTER 4

## "DEVELOPMENT" CHALLENGED

Ever since Aristotle tried to unravel the process by which living things evolved from stage to stage, men have searched for the causes and consequences of "development." Behind the search has been a deep-seated assumption that living things proceed from a less sophisticated state to a more sophisticated state, all the time increasing their abilities to fulfill their inherent potentials. It has been an assumption filled with hope. It also has been the source of frustration, arrogance, and humiliation. For the notion of development presumes that there is some target, some higher plane of existence which individuals or nations can achieve. Those who fail to reach that point have cheated their own potential; they are "underdeveloped." To make that distinction between things that were

**111**

developed versus those that were underdeveloped there had to be, of course, some agreement about the character of the target. Thus, "development" became both a measurement and an ideological goal.

Decolonization after World War II stimulated new concern over the means and effects of socioeconomic changes. The rationale for colonization had included a strong presumption of development. European, American, Chinese, Japanese, and former American slaves settling in Liberia all couched their colonial occupations in terms of bringing "civilization" to benighted peoples. Since 1950 the concept of development has become a more explicit guide to policy. Cultural biases and myths remained but were now harder to discern. It looked as though there would be a universal consensus on the target for change, though there were still disputes over the best strategies for reaching that destination.

What the development specialists chose to measure reflected the spreading consensus. They measured a country's gross national product, the sum of goods and services produced annually. They also measured nonagrarian employment, urban population, and literacy, all deemed crucial for raising a country's GNP. "Per capita energy consumption" was still another measuring rod of development, since economic growth depended on generating sufficient energy to support an industrial "infrastructure"—i.e., that network of communication and transportation facilities that became the skeleton for any modern society. World maps indeed showed that the low-energy-consumption areas coincided neatly with the boundaries of Third World nations.[1] Measuring development success by kilowatts couldn't help but dampen any protests against ecologically questionable hydroelectric projects such as the huge dams in Egypt, India, or Cambodia.

---

1. David Hilling, "Open Roads for Exports and Imports: The Third World," *Geographical Magazine* (UK), October 1973, p. 29.

It became difficult to tell whether the objects of measurement were mere indicators of some loftier abstract goals or were the goals themselves. Industrialization initially was seen as a means for achieving societal well-being. Soon industrialization *was* well-being. Socialists, capitalists, and communists agreed that in industrialization and economic growth lay the hope of Third World nations.

Behind the urgent efforts to promote economic growth in the postwar decades lay a motley set of motivations. Some experts and politicians were genuinely troubled about the legacy of colonialism and the global gap between rich and poor. Other development specialists were anxious to try out new skills and theories. Underdeveloped countries were for them intriguing "problems" to be "solved."

Governments supporting development programs in the Third World were no less mixed in their motives. The Great Powers viewed the Third World as an arena in which international competitions for allies and resources would be played out. Development aid might alleviate poverty and injustice, but it would also serve as a vehicle for building valuable buffers and alliances. Those industrialized nations that came out of World War II in defeat or shorn of former empires could use development assistance to revive their international influence and to ensure markets.

Politicians and bureaucrats in the ex-colonies accepted aid from one or usually several industrialized countries and from international agencies for a similarly mixed bag of reasons. Many Third World leaders had been trained in North America, Europe, or the Soviet Union and returned home convinced of the validity of pro-growth, pro-industry theories propagated abroad. Policy makers in the developing countries also were nervous about the aspirations of their own countrymen, aspirations that were bound to outstrip local resources and to threaten the governments in which they held authority. Many of these same leaders remembered, moreover, the humiliation of colonialism and believed that only through national economic expansion could their nations hope to fend off new forms of foreign domination.

After almost a generation spent in pursuit of industrial growth, policy makers and administrators in donor as well as recipient countries began to voice doubts. The UN is launching the Second Development Decade in 1975. On its eve there are serious reassessments of the meaning and consequences of development. The First Development Decade began with a fanfare; the second begins with a question.

There are four principal challenges now to the standard postwar notion of development:

1. Development when defined as growth and industrialization produces unpredicted environmental costs. In the cities this takes the form of polluted air; in the countryside the "Green Revolution" fosters reliance on petrochemicals, while huge dams produce soil erosion and damage to fish.[2]

---

2. The debate over the negative effects of the so-called Green Revolution is carried on in Robert Katz, *A Giant in the Earth* (New York: Stein & Day, 1973); Melvin J. Grayson and Thomas R. Shepard, *The Disaster Lobby* (Chicago: Follett, 1973); Richard W. Franke, "Miracle Seeds and Shattered Dreams in Java," *Challenge* 17, No. 3 (July/August 1974): 41–47; Vance Bourjaily, "One of the Green Revolution Boys," *Atlantic*, February 1973, pp. 66–76; Brian Phelan, "Harvest of Rhetoric," *Far Eastern Economic Review*, 16 April 1973, p. 48; Richard Weintraub, "Food . . . The Diminishing Necessity," *Washington Post*, 10 February 1974.

Concerning the environmental effects of dams in the Third World, see John E. Bardach, "There Is More to Dams than Meets the Eye," *Asia*, no. 20 (Winter 1971): 9–30; Georgiana G. Stephens, "Egypt," *Atlantic*, January 1972, pp. 12–16; Claire Sterling, "Superdams: The Perils of Progress," *Atlantic*, June 1972, pp. 35–41; D. E. Ronk, "A Dam for All Seasons," *Far Eastern Economic Review*, 3 April 1971, pp. 124–25.

2. Development has put too much stress on the need for urban growth, to the point that urbanization has outrun the available jobs, housing, and services and undermined the urban fabric of life.[3]

3. Development has not spread its benefits evenly. There are widening gaps between rich and poor within Third World countries.[4]

4. Third World development may be raising African and Asian GNPs, but the gap between affluent and developing countries is still widening, not narrowing.

5. Development as currently conceived is entangling recently decolonized nations in new webs of dependency. Their citizens are learning preferences for foreign styles; their economies rely on influxes of foreign capital and technology; their currencies are affected by an international monetary system they cannot control.[5]

The environmental costs of unrestrained economic growth have been a principal factor prompting skeptical rethinking about development. But the new skepticism has not produced wholehearted backing for the environ-

3. Robert McNamara, head of the World Bank, has been a leading advocate of reversing emphasis on urban expansion. For a critique of the extent to which the World Bank's loan practices have actually implemented this reversal, see *New York Times*, 8 June 1973.

4. A study by Irma Adelman and Cynthia Taft Morris found that among 74 underdeveloped countries between 1957–68 there had been (a) *decrease* in political participation, (b) a decrease in the relative share that the poor had in the total national income, and (c) frequently even a *decrease* in the poor's well-being in absolute terms. Irma Adelman and Cynthia Taft Morris, *Economic Growth and Social Equity in Developing Countries* (Stanford: Stanford University Press, 1973).

5. Among the most provocative reassessments of development theories are Robert A. Packenham's *Liberal America and the Third World: Political Development Ideas in Foreign Aid and Social Science* (Princeton, N.J.: Princeton University Press, 1973). Also refer to Samuel Huntington, "The Change to Change: Modernization Development and Politics," *Comparative Politics* 3 (April 1971): 283–322.

mental movement in the Third World. In fact, the web of
issues in which environmental deterioration now exists in
Asia, Africa, and Latin America has generated ambivalence at best and outright hostility at worst toward
Western environmentalists. For the disappointment in
conventional strategies derives in large part not from
their economic excesses, but from their economic short-
falls, especially their inability to close the domestic and
global chasms between rich and poor.

Responding to the ambivalence concerning the environmental issue, the United Nations launched the Second
Development Decade by commissioning a long-range
study to determine precisely how great would be the
impact of environmental controls on economic progress
over the next seventy-five years. Financed largely by the
Dutch government and headed by Nobel prize-winning
economist Wassily Leontief, the UN study will make
consumption, production, and environmental projections
for the years 1980, 2000, and 2050. Leontief and his
colleagues will question whether environmental preservation inevitably retards Third World economic advancement.[6] At the same time, however, UN officials continue
to formulate plans that call for a 6 percent annual growth
rate for Third World countries and continue to use the
GNP as their yardstick for measuring development success.[7]

When environmental issues do break through this
justified Third World mood of wariness, they are marked
by characteristics distinctive of their underdeveloped
setting. First, environmental issues are likely to be less
urban since most Third World countries remain largely
agrarian. Second, and reflecting this rural thrust, environmental questions will quickly spill over into problems that

6. "World Environment Newsletter," World, 17 July 1973, p. 38.
7. Peter Cook, "The Development Hangover," Far Eastern Economic Review, 16 October 1971, pp. 28–31.

are not specifically man-made, such as drought or flooding. Third, the problems of the physical environment will be discussed in the context of the debate over population growth and control. For the high birthrates that have thwarted so many ambitious five-year plans make the choices between environmental risk and rewards of growth even more cruel than in industrialized nations where population rates are declining. Fourth, the ways in which the environmental issues are tackled will be determined by certain political aspects of underdevelopment. These include limited avenues for popular participation, an administrative orientation toward politics due in part to the style of politics inherited from colonial rulers and a severe scarcity of governmental resources. Finally, policy makers in underdeveloped countries will be especially vulnerable to external pressures as they weigh environmental options.

### POLITICAL LIMITATIONS
### OF UNDERDEVELOPMENT

Political underdevelopment refers to a system's inability to control its own universe adequately enough so that it can formulate and implement meaningful decisions. Intrusions from foreign nations limit a developing country's efforts to make policy. In addition, the government of a politically underdeveloped country faces fluctuating international markets and diplomatic negotiations that it must constantly respond to without being able to direct.

The Mideast war of 1973 which drove up petroleum and fertilizer prices reminded one that to be underdeveloped is not just to be poor, but to lack sufficient power so as to make the decisions that would lessen that poverty. Third, political underdevelopment is an inadequacy of public resources with which to carry out decisions and to multiply and expand policy options. A handful of nationals in top ministerial posts who possess university degrees can conjure up all the *possible* policy approaches to the problems of offshore oil or wildlife preservation. But to be effective a government needs revenues and skilled manpower down through the lower bureaucratic ranks so that possible options are in fact real options. Third World governments are limited by their small tax bases and inefficient tax collections, by the strings attached to foreign aid, by the educational systems that measure success in terms of *primary* school graduates, and by the overseas brain drain which lures Egyptian, Nigerian, Indian university graduates to stay on in the United States or Europe rather than return home where their skills are more needed but their salaries are lower.

In environmental affairs, political underdevelopment is, at bottom, foreign dependency plus too few policy options. For example, imagine the problems that the banning of DDT has posed for Third World policy makers. You are an agricultural policy maker in Ghana and receive a report that Britain has banned the insecticide DDT. Your own agricultural plans depend on pesticides to increase Ghanaian rural production and thus you are reluctant to withdraw DDT without sound evidence. You cannot assume that Ghana's ecological conditions are identical to the UK's and therefore you cannot automatically follow the British lead and ban DDT in Ghana. On the other hand, your administrative personnel lack the necessary scientific staff and data sources with which to evaluate DDT's harm in Ghana. Moreover, being an experienced governmental official, you know that there were probably some important nonscientific factors which played a part in the British policy decision. Yet political participation in Ghana is not yet institutionalized

enough to provide policy advice on many specific issues
of this sort. This quandary confronting the Ghanaian
agricultural planner is just as symbolic of political underdevelopment as are army coups or urban riots.[8]

The administrative dimension of environmental politics also underscores some of the special difficulties borne
by Third World governments. The bureaucratic pitfalls
between decision and implementation affect Britain as
well as Ghana. But a mark of underdevelopment is a
bureaucracy plagued by inefficient procedures, structures
that are not goal-oriented, recruitment that favors status
or personal connections over functional skills.

There is the incidence of corruption too. Many Third
World policy makers are annoyed when foreign experts
refuse to treat the problems of corruption seriously. Since
bribery can destroy effectiveness and credibility, it can
undermine administrative operations as assuredly as can
the lack of engineers or statisticians. There is also the
difficulty of obtaining feedback in developing country
administration. Effective environmental policies require
means to monitor consequences so that original policy
can be reinforced, modified, or abandoned. Obstacles to
interest-group formation, citizen participation, and scientific communication reduce the amounts and reliability of
feedback information that regimes can use to refine their
environmental policies.[9]

Examining several subissues in detail can clarify the
conditions which distinguish environmental politics in
developing nations. The two subissues to be discussed are
(1) tourism and (2) oil exploration.

---

8. This illustration was suggested to the author by Ian Burton,
professor of geography, University of Toronto, during a seminar on
Man-Made Hazards and the Dissemination of Scientific Information,
Clark University, Worcester, Mass., 15–18 January 1974.

9. Thomas B. Smith, "The Study of Policy–making in Developing
Nations," *Policy Studies Journal* 1, no. 4 (Summer 1973): 244–49.

## TOURISM

The very widening of the gap between Third World and industrial nations has promoted tourism. For as citizens of industrialized countries find their own territories eaten up by commercial utilization and possess more time and funds for leisure, they look around the world for those "unspoiled" havens where the air and land are pollution-free and life is reminiscent of their own allegedly "simple" past. The host countries that court German, Japanese, or American tourists are eager to build up foreign currency reserves and to promote enterprises like tourism that lessen dependence on their traditional raw material exports. But tourism involves a double irony. The tourists, by seeking out these spots, are themselves adding to their despoilation; the governments whose tourist bureaus are attracting the foreign influx use that income to hasten their country's entrance into the industrial age.

Because tourism is becoming a major ingredient in Third World economies and because that tourism is overwhelmingly foreign in character, the politics of regulating it for the sake of the environment are filled with the impossible choices and frustrating limitations that so distinguish Third World politics in general. The chief participants in debates over the costs and benefits of tourism are government civil servants and cabinet officers, especially those in the Ministries of Economic Development and of Tourism on the one hand, and foreigner-dominated interest groups on the other hand. The domestic citizenry may be taken into account in the policy discussions, but rarely are they actors in the process. The alignments do not always pit the foreigners versus the locals; at times the environmental advocates may include a combination of locals and foreigners. Options available to local decision makers are severely limited by the pressing need to promote economic development and by dependency on foreign assistance. To multiply policy options may require an overseas donor willing to put up the new funds.

In sub-Saharan Africa the controversy over tourism has revolved around the question of wildlife preservation. Colonial administrations in countries such as Tanzania and Kenya had established game reserves which, after independence, became major tourist attractions and sources of vitally needed foreign currency. By 1973, tourism had become Kenya's number-one earner of foreign exchange. In 1971 alone, some 400,000 tourists visited Kenya, spending a total of $67.5 million. In neighboring Tanzania tourism earns only one-fifth that much, but the income is steadily rising.[10] Seeing African wildlife is considered the prime motivation for foreigners coming. Africans who might have the resources to be tourists are urban dwellers and look to Europe as the place to travel if possible. One governmental policy that might transform the politics of tourism and environment in countries like Tanzania is official encouragement of urban Africans to take a greater interest in their own habitats. Today, though, it is foreign wildlife preservation organizations that act as the principal pressure groups in wildlife politics. American and West German groups are especially vocal. Furthermore, the tourist companies that have the most to gain from the preservation of the parks and reserves are also in the hands of Europeans and Americans. In Kenya, for instance, there are eighty licensed hunter-guides. All are European: "The tradition of white hunters is so ingrained that the few Africans or Asians training for the profession will say, 'I am becom-

---

10. Roger M. Williams, "The Politics of Wild Animals," *World*, 13 February 1973, p. 25.

ing an 'African White Hunter' or 'an Asian White Hunter.' " [11] And as of March 1973, only one safari firm in Kenya was totally African-owned.[12]

Pressured by overseas interest groups, conscious of their growing economic stake in the tourist trade, and increasingly anxious to protect their own environment out of nationalist pride, several African regimes have taken steps to safeguard their wildlife. The Kenyan government of Jomo Kenyatta took an active part in the 1973 international wildlife conference and helped to draft a proposal that the sixty-one participating governments would sign establishing a permit system covering both the export and import of endangered species. Reflecting the international character of tourism politics, Kenya's partners in the drafting were the environmental affairs office of the U.S. State Department and the International Union for the Conservation of Nature and Natural Resources, headquartered in Geneva.[13] Again drawing on international resources, the Kenyan government in 1973 signed an agreement with the New York Zoological Society for joint financing of a water pipeline system to keep the Masai tribe's cattle from damaging a nearby 150-square-mile wildlife sanctuary. The society and the Kenyan government each pledged $140,000 to start the project.[14]

For its part, Tanzania, under the regime of Julius Nyerere, instituted eight new game sanctuaries during its twelve years of independence. Even more significant given the limited budgetary resources plaguing developing countries, Tanzania has been pouring 3 percent of its annual budget into conservation. This amounts to *three times* more per capita than the United States allots

---

11. Francine du Plessix Gray, "On Safari," *New York Review of Books,* 28 June 1973, p. 26.
12. Ibid.
13. *New York Times,* 11 February 1973.
14. *New York Times,* 12 March 1973.

for similar programs. In addition, Tanzania has created a Wildlife College at Moshi, supported largely by funds from the United States, and has promoted wildlife educational programs in the primary and secondary schools.[15] The Tanzanian regime is credited by some observers as being the most wildlife-conscious of African regimes, though reductions in British foreign subsidies for its parks has intensified the budgetary problems.[16]

Difficulties arising from the preservation of wildlife are compounded by being interwoven with the politically sensitive question of land distribution. For in Africa, as in many developing countries, land is the principal source of wealth and status. Land is thereby a major object of political competition and conflict. The wildlife sanctuaries may attract tourist dollars and protect endangered species, but they also take valuable land out of the domestic market. Many African tribes, such as the Masai, need large tracks of land to carry on their traditional livestock practices. Other tribal or ethnic groups press governments to open up lands for new agrarian cultivation. An estimated 50 percent of East Africa's land is deemed unsuited for farming.[17] Yet populations are growing, and governments are becoming increasingly attracted to development strategies that stress rural development and curb unplanned urban expansion. Kenya has a 3.3 percent annual population growth, one of the highest in the world; virtually all its tillable land is already in use; and there has been increasing conflict between the dominant Kikuyu tribe and other ethnic groups for farm land. This is not a situation in which allotment of budgetary funds for wildlife reserves can win much domestic popularity for a politician. Moreover, African nationalists are wary of the

---

15. Gray, "On Safari," p. 28.
16. Cyril Connolly, "Carnage in Eden," *New York Review of Books*, 25 January 1973, p. 19.
17. Williams, "Politics of Wild Animals," p. 26.

tourist industry. It may bring needed currency, but it also celebrates the gap between rich whites and poor non-whites and portrays developing nations as mere playgrounds for the industrialized countries. In Caribbean countries such as Jamaica, Trinidad, and Puerto Rico, too, there are protests against becoming merely "nations of busboys." [18]

In Africa and the Caribbean nations the environmental issue, whether it takes the form of wildlife preservation or protection of unspoiled beaches, can become politically unpalatable if it: (1) seems to benefit only foreign economic interests, (2) has little relevance for a local population whose values are being oriented more toward urban rewards, and (3) is likely to reduce the land available to growing local populations for their own commercial advancement. By contrast, environmental protection is likely to be *most* politically acceptable when: (1) the costs are shared in part by overseas groups, (2) local populations merge nationalist sentiments with a desire to protect their natural surroundings, (3) tourism is so controlled that the stigmas of racism and imperialism are minimized, (4) overseas tourist interests themselves see environmental protection in their long-run economic advantage even if it imposes short-run economic restrictions, and (5) population control programs and farming improvements can reduce the pressure for exploitation of available land.

## OIL EXPLORATION

On the surface, the fuel crisis of the 1970s has been an unqualified boon to a number of developing countries.

---

18. See John Bryden, *Tourism and Development: A Case Study of the Commonwealth Caribbean* (New York: Cambridge University Press, 1973).

But not all. For nations such as South Korea or India which are on the verge of agricultural and industrial breakthroughs, the shortages meant exorbitant prices and inadequate supplies, causing turmoil for their economic planners. Those developing countries for whom fuel shortages did present a golden opportunity were those with oil reserves. Among those nations were Nigeria, Indonesia, Malaysia, Algeria, Thailand, Peru, Taiwan, and, of course, the otherwise undeveloped nations of the Middle East, especially Saudi Arabia. In addition, Singapore and Caribbean nations or territories such as Trinidad and the Dutch Antilles stand to gain as sites for now expanding petroleum refining facilities.

A number of issues are gaining political attention in these countries, thanks to this new drive for oil, but environmental questions rank low on most lists. The sort of political controversy that swirls around oil drilling off California's shores or proposed refinery construction off New Hampshire's coast is rare indeed in the Third World. Instead, the issues spun off oil concern: (1) governmental control of petroleum exploration, (2) royalty rates, and (3) the number of jobs that the local population will gain from any new oil investments. They are issues which lie at the heart of contemporary underdevelopment: national sovereignty and economic growth. Environmental questions are luxuries. Third World governments are more worried about how they are going to pay for their huge imports and how they are going to generate jobs for increasing numbers of young school-leavers.

Environmental concerns about oil spills *can* demand attention when the petroleum industry does not create more local jobs, when oil threatens an existing commercial sector such as tourism, or when oil investments curtail national independence. For example, in the Virgin Islands, Puerto Rico, and Trinidad the same government critics who assert that the governments' friendly welcome to oil companies has not relieved serious unemployment and has compromised national integrity also have argued that oil exploration and refining is endangering one of the few things that local islanders can really call their own, their natural environment.[19] Moreover, oil spills could threaten the territories' other major job sector, tourism. The president of the Caribbean Conservation Association reminded that, "Tourism is our basic economy. It hangs on clean water and beaches, sports, fishing and sailing. But a wide area of the Caribbean is headed for environmental disaster." He continued, "Saint Croix gets oil spills two or three times a week when tankers hit submerged objects or crash into rocks. The big spills get the publicity, but it's the day-to-day seepage that is ruining our waters. You might call it death on the installment plan." [20]

In few developing countries, however, have these conditions been blatant enough to make environmental disruption a serious political issue. Nigeria is more typical. For Nigeria the global escalation of oil demands came at a time when its own resources were being exploited more fully than ever before. More important, it came at the end of the Biafran civil war. The war-drained central military government of General Gowon needed great infusions of foreign currency to pay for postwar recovery and to sustain its war-inflated national army which could not be demobilized because of job scarcity in the civilian sector. One reason why the Gowon regime fought so fiercely against the Ibo-led rebels of Biafra was that the secessionist state had claimed for itself Port Harcourt, the site of the most promising oil explorations. With secession quashed, there was still the problem of distributing

---

19. See, for instance, Allen Harris, "Shell Oil Barges into Caron Swamp," *Tapia*, 10 June 1973, p. 3. *Tapia* is a weekly paper published by Trinidad's radical-reformist New World movement.
20. Quoted in the *New York Times*, 28 January 1973.

the oil revenues among the fragile federal system's twelve states. Should distribution be according to population or some other criterion of need? Here was a political issue that was more pressing than abstract notions of coastal preservation. Oil potential also posed diplomatic problems that had to claim the government's attention: Should Nigeria break off relations that it had enjoyed with Israel? Should it support the Arab oil boycott, which of course gave Nigerian oil higher value on the world market?

Thus oil has indeed generated critical issues in Nigerian politics. Yet environmental control has not been a salient matter. With one indirect exception. Nigerian oil happens to possess one of the lowest sulphur contents in the world. Air pollution issue saliency in industrialized countries has made low sulphur crude more valuable than ever. Arab oil and most North American oil have high sulphur ratings and are more restricted by new air-quality controls imposed in Europe and the United States. Thus, while pollution is scarcely a political issue in Nigeria, the prominence of that issue in its potential export markets has vastly increased its oil's economic value. A partnership between the Nigerian Oil Company and an industrial combine of British Petroleum–Shell made it possible to launch an ambitious reconstruction and development plan of $5 billion and enable it to eliminate most of its previous national debt of $1.5 billion.[21] Whereas in 1965, on the eve of the civil war, Nigerian wells produced only 120,000 barrels per day, by 1973 they were yielding 2 million barrels per day, which amounted to revenues equalling 11 percent of the gross national product.[22]

In the Southeast Asian countries of Malaysia and Indonesia as well, conditions have made the environmen-

21. "Africa's Oil-Producing Countries," *African Progress*, December 1973, p. 32.
22. Ernest J. Wilson, "Energy, Africa and World Politics," *The Review of Black Political Economy* 3, no. 4 (Summer 1973): 31.

tal implications of oil exploration a nonissue. Malaysia's Alliance party government is trying to hold together a country composed of three mutually suspicious ethnic groups (Malays, Chinese, and Indian). In addition, Malaysia is a federal nation with two of its states, Sabah and Sarawak, separated from the peninsula by hundreds of miles of water, economic backwardness, and even greater ethnic complexity. Its principal strategy for sustaining its own primacy and maintaining national unity is to guarantee rapid enough economic growth that all groups will feel that they are getting a piece of an expanding pie.[23] While Malaysian oil discoveries are modest, they are supplying the regime with the politically crucial revenues with which to carry out its strategy. Sabah's 29 percent growth in GNP between 1972–73 and Sarawak's 14.1 percent GNP rise in the same period have been heralded by the central government as a reflection not only of its own competent planning but also of Malaysia's promising future in oil production.[24] The decline of Malaysian unemployment has been due in part to the government's aggressive recruitment of foreign industries to the peninsular states, and among those considered most promising in another oil-related field, petrochemicals.[25]

Despite the low interest expressed in environmental issues in Malaysia, the Alliance regime, priding itself on

23. Philip Bowring, "Malaysia: The Cycle to Success," *Far Eastern Economic Review*, 27 May 1974, pp. 55–61.

24. Federation of Malaysia, Ministry of Foreign Affairs, *Malaysian Digest*, 15 January 1974, pp. 6–7.

25. Ibid., p. 2. See also *Malaysian Digest*, November/December 1973, pp. 5–6. Following the 1974 national election the victorious Alliance leadership reshuffled the cabinet giving responsibility for environmental affairs to a Chinese politician, making him head of the reorganized Local Government and Environment Ministry. Assigning a Chinese to the environmental ministry is possibly an indication that it is not deemed crucial politically, for the government is dominated by Malays now and they have kept the most powerful minorities in their own hands. M. G. G. Pillai, "Upsetting the Neighbors," *Far Eastern Economic Review*, 20 September 1974, p. 13.

nonideological pragmatism, has acknowledged the negative by-products of the country's current boom. In 1974, in the midst of economic euphoria, the government established an Environmental Quality Council with authority to license water-disposing firms. The regime also initiated fines up to $25,000 for oil discharges in Malaysian coastal waters.[26] The latter is a sign of the government's sensitivity to challenges to its sovereignty over the Malacca Straits, a major route for supertankers carrying Arab oil to Japan. The running aground of a Japanese supertanker early in 1975 and its discharge of thousands of tons of oil into the Straits only heightened this sensitivity. Again, nationalist sentiment can award political attention, if not necessarily policy priority, to environmental protection.

In Indonesia, as in Nigeria, the crude oil has a low sulphur content which gives it special attractiveness in American and European markets that must adjust to stricter air pollution controls. Similarly, the army regime in Indonesia is hoping that oil revenues will patch over dangerous domestic political cleavages. After the fall of the Sukarno regime and the nationwide purge of suspected communists, the military under General Suharto focused its energies on economic rationalization and expansion. Oil is vital to that political strategy. Sensitive to mobilized nationalist sentiments, the government sought to defuse the charge of "sellout" by strictly regulating the operations of American, European, and Japanese oil investors. The organization that draws up and supervises these contractual arrangements, Pertamina, has become a model that other Third World governments are looking at with hopes of emulating. Leftist critics acknowledge that foreign companies such as Atlantic Richfield, Caltex (Texaco and Standard Oil of

---

26. "World Environment Newsletter," *Saturday Review/World*, 1 June 1974, p. 36.

California), and Union Oil have had to make substantial concessions to Pertamina. The government corporation keeps owernship of the oil and allows drilling companies to keep up to 40 percent of what they extract. Pertamina —called a "state within a state"—in turn spends sizable amounts on private hotels and private planes as well as on such projects as an employee hospital. By 1973 Pertamina's income accounted for a full 50 percent of the Indonesian government's annual revenues.[27]

According to their critics, Pertamina and Suharto's Western-trained advisers are too friendly to overseas businessmen and too eager to accept foreign investment as the bulwark of national development. They claim that there is inadequate "trickle down" from Pertamina's oil ventures to the 125 million Indonesians who live on a per capita income of $85 per year and who migrate by the thousands to cities that lack both jobs and shelter to accommodate them. There is to date no indication that these arguments have become mobilizing points in current Indonesian politics to such a degree that oil exploration will lose its appeal for central policy makers or trigger an environmental movement in that country.

An examination of the politics of oil in the Third World oil-producing countries suggests additional generalizations regarding environmental issue creation. First, environmental concerns are secondary to economic needs so long as "development" is defined by technical assistants and local policy makers largely in terms of industrial

27. Joseph Lelyveld, "Americans in South Asia Join in Scramble for Riches," New York Times, 24 June 1974. Long-time Southeast Asia correspondent Robert Shaplen reported from Indonesia in 1974 that the head of Pertamina, General Ibu Sutowo, was purported to be "perhaps the richest, and certainly one of the most powerful and most controversial" government officials in the entire region. A government anticorruption commission report singled out Pertamina for special investigation. Robert Shaplen, "Letter from Indonesia," New Yorker, 1 April 1974, p. 66. See also Crocker Snow, Jr., "Indonesia Gains Far East Strength Because of Middle East Oil Policy," Boston Globe, 22 November 1973.

expansion and rising GNP. Second, those regimes nervous about domestic conflict are likely to try to heal wounds with money and be least likely to treat seriously alleged environmental dangers from oil expansion. Third, where other industries on which the developing countries depend—especially tourism—are not directly affected by oil exploration, the environmental issue stands less chance of gaining saliency. Fourth, perhaps the most likely way for environmental issues to capture politicians' attention is for them to ride piggyback on other issues, especially the issue of foreign exploitation or overseas control.

## NEOIMPERIALISM AND POLLUTION HAVENS

The dilemma of underdevelopment is choosing between two unwanted or compromising conditions. A government that wants economic growth may feel that it must accept foreign intervention and potential environmental hazards. The very opportunity for attracting overseas capital may come from an affluent nation protecting its own environment. As affluent countries enact stricter environmental legislation, their businesses look to less developed nations as sites for factories. True or False:

—It is immoral for industrialized countries to "export pollution"?
—Is it immoral for leaders of less developed countries to establish "pollution havens"?
—The United Nations must establish uniform pollution standards for agriculture and industry, adhered to by all countries of the world?[28]

---

28. "Environmental Newsletter," *World*, 19 December 1972, p. 45.

This is not the sort of quiz that a policy maker enjoys. But it capsulizes the environmental dilemma of underdevelopment. As one might expect, Third World politicians are divided in their answers. Imposition of universal standards might solve the dilemma but in so doing it could harden the gap between rich and poor. As an Indian official at a Puerto Rico preparatory meeting before the Stockholm Conference quipped, "The wealthy worry about car fumes. We worry about starvation." Another delegate at the same conference, an engineer from the Barbados Ministry of Health, noted that even for Third World policy makers to imagine what a "pollution haven" is requires more environmental awareness than most had yet acquired: "Developing countries always run the risk of being pollution havens for the kinds of industries and projects the advanced countries, with experience, no longer want. It is hard for us to turn them away when our own needs are so great. We need the courage of our convictions. But first we have to develop our convictions." [29]

The debate over standards that would apply to Third World as well as affluent countries, therefore, does not present a neat alignment of poor versus rich. There are rich and poor each side. In fact, because of the differing bureaucratic outlooks and interests mentioned earlier, there will probably be opposing stands even within a single government. The Barbadian Health Ministry official may find himself arguing not only with an official of Royal Dutch Shell, but with a fellow Barbadian who represents the Ministry of Economic Development. Furthermore, even those officials who are wary of their nations becoming pollution havens are suspicious of any universal standards imposed from outside. They seem to be just one more intervention, one more hole in the wall of national

---

29. Irwin Goodwin, "Pollution Lessons for Poor Nations," *Trinidad Guardian*, 4 September 1971.

sovereignty. The ambassador of Sri Lanka (formerly Ceylon) voiced such suspicions, warning his Third World colleagues not to adopt views just because they were being pressed by larger or richer governments:

> The governments of developing countries, their economists and planners must not and will not allow themselves to be distracted from the imperatives of economic development and growth by the illusory dream of an atmosphere free from smoke or a landscape innocent of chimney stacks. We must not, generally speaking, allow our concern for the environment to develop into a hysteria.[30]

Where "pollution haven" becomes a meaningful concept and arouses concern is likely to be: (1) where the hazardous consequences of a new industry obvious to the affected public; (2) where those conditions are easily traced to a foreign company; (3) where nationalist sentiment on a broader scale has been mounting and can serve as a glass through which to perceive the hazard and its cause. For instance, as anti-Japanese feeling has spread in Southeast Asia, where memories of harsh Japanese occupations during World War II still linger and are fueled by resentment against current massive Japanese investments, "pollution haven" has gained some political significance. In Thailand, a country sensitive enough of its sovereignty that it never was colonized, there have been protests against the pollution hazards from Japanese-backed petrochemical complexes.[31] This awareness may in turn arouse anger about other alleged pollution sources owned by persons who either are foreigners or who, as in the case of the Chinese of Thailand, are not accepted as full members of the nation. As Thai students were leading

---

30. Quoted in L. Berry and R. W. Kates, "Environmental Problems —A View from East Africa" (Manuscript, 1973), p. 16.
31. Gemini News Service (UK), in *Guyana Graphic*, 12 June 1972.

protests against Japanese businesses, they were also demanding forceful action from the minister of industry against sugarmill owners, largely Thai Chinese, whose dumping of untreated waste was threatening to pollute a popular resort area.[32]

Industrialized nations and their interest groups disagree among themselves regarding pollution havens and possible worldwide standards. The U.S. Environmental Protection Agency, for instance, was the sponsor of that pre-Stockholm Puerto Rico meeting. The EPA tried to persuade Third World delegates that they should modify their economic priorities for the sake of their own environments. Likewise, in Japan local critics of the ruling party see pollution havens as one more harmful effect of the regime's all-out drive for industrial primacy in international arenas. Businessmen likewise are apparently divided in their opinions. Obviously many see the unevenness of environmental standards around the world, and especially the relative lack of restrictions in underdeveloped areas, as posing an attractive alternative to shouldering the added costs of meeting new pollution criteria at home and to fending off mounting domestic protest. On the other hand, business executives in companies with widespread international trading networks view the growing unevenness in environmental laws as a severe handicap to their operations. Instead of trying to cope with laws that impose a variety of restrictions using a variety of standards, these business policy makers would prefer global uniformity so that they could standardize their own operations. Car manufacturers, detergent makers, petrochemical companies, and others face increasingly varied international markets due to uneven environmental political development.[33]

---

32. "Letter from Bangkok," *Far Eastern Economic Review*, 2 July 1972, p. 50. See also, "Environmental Newsletter," *Saturday Review/World*, 6 November 1973, p. 39.

33. *Wall Street Journal*, 29 November 1971.

Brazil has been perhaps the most outspoken Third World country in its contempt for the notion of pollution havens. Between 1968 and 1973 Brazil grew faster economically than any other country in the world, with a growth rate that never dropped below 9 percent. With some commentators projecting that by 1990 Brazil will replace Britain as a major industrial power, it is looked to by many Third World planners and by potential investors as "the new Japan." [34] Its formula for this rapid economic growth has been summed up as follows:

1. *Abolition of politicians.* The military has ruled with the assistance of technocrats since 1964.[35]
2. *Welcoming foreign investors.* Foreigners may not remit more than 36 percent of their capital in any three-year period, but there has been no reluctance to allow the car industry to become 100 percent foreign-owned, tobacco 91 percent, rubber 82 percent, and chemicals 54 percent.
3. *High internal savings.* Brazilians are reinvesting about 22 percent of their GNP each year, due first to an effective computerized tax collecting system and, second, to a policy of incentives for saving.
4. *Regular devaluations of the currency.* The government keeps devaluing in order to maintain the international trading attractiveness of Brazilian goods.
5. *Domestic stability.* Dissent has been quashed

---

34. "The Next Japan Emerges from the Jungle," *Economist* (London), 15 December 1973, p. 85. Interestingly enough, at the same time that Brazil is being labeled the "new Japan," Nigeria is being labeled the "next Brazil."

35. One of the most perceptive books on Brazilian politics is Alfred Stepan, ed., *Authoritarian Brazil* (New Haven: Yale University Press, 1973). For premilitary Brazilian politics, see John D. Wirth, *The Politics of Brazilian Development, 1930–1954* (Stanford: Stanford University Press, 1972).

leaving parties and trade unions ineffective and offering little outlet for dissatisfaction with the increasingly unequal distribution of Brazil's economic benefits.

6. *High tariff walls.* Legislation prevents public enterprises from importing goods available from domestic manufacturers.

7. *Belief in free enterprise.* The government controls public utilities, part of the steel industry, and all of the oil industry; but most of the economy is in private hands, with an estimated one Brazilian in three self-employed and would-be "Rockefellers" given free rein.[36]

Integral to this growth formula has been the official tolerance of environmental disruption in both the countryside and the city. Environmentalists, along with anthropologists concerned about the indigenous Indian population, have criticized the military rulers for their unrestrained development of the Amazonian interior. Preparing for the World Population Conference in Bucharest in August 1974, for instance, the Brazilian delegation announced that it would reject population-control plans. Instead, Brazil's regime planned to *increase* the country's present 104 million citizens to 200 million by the 1990s in order to populate its vast interior and to expand its domestic consumer markets.[37] At the same time, a leading Brazilian conservationist, José Piquet Carneiro, resigned as director of the official Brazilian Foundation for

---

36. "The Next Japan Emerges from the Jungle," pp. 85–86. During the 1960s the richest 5 percent of Brazil's population increased its share of the GNP from 44 percent to 50 per cent, while "the poorest 80 percent saw their share of the GNP drop from 35 to 27.5 percent." E. Bradford Burns, "Brazil: The Imitative Society," *Nation*, 10 July 1972, p. 18. For a similar analysis, see Fred Halliday and Maxine Molyneaux, "Brazil: The Underside of the Miracle," *Ramparts*, April 1974, pp. 14–20.
37. *New York Times*, 9 June 1974.

the Conservation of Nature claiming that the government was "turning the Amazon into a desert, destroying its forests, rivers and animal life, and there was nothing I could do to stop them." He was frustrated by a regime that "thought progress was cutting down trees and building roads." [38]

In urban areas there has been until recently no legislation ensuring that services and codes matched industrial and population expansion. São Paulo, Brazil's and Latin America's largest city with 6 million people, was selected by Volkswagen in 1957 to be the site for its manufacturing operation and has since become the focus of foreign investment in Brazil. At the same time, half the homes in São Paulo do not have running water. The city is noted for its congestion and its high level of air pollution. Only on the eve of the city's 419th anniversary did the mayor and municipal council approve a "Plan for Integral Development" which included strict antipollution measures and a meaningful building code.[39] Not long after this pronouncement, another commentator described São Paulo as a city "on the verge of a breakdown." [40] It will take an energetic government to protect an urban environment in which there are more automobiles than in any city outside Los Angeles, in which there are 24,000 factories and an estimated 300,000 poor migrant squatters. "On any given day a pedestrian in downtown São Paulo inhales more than three times the pollutants found in Chicago." [41]

Even after the UN's Stockholm Conference the Brazilian generals and technocrats appeared to be unalarmed at the long-term costs of this unrestrained economic drive. At Stockholm the Brazilian delegation accused the

38. Quoted in *New York Times*, 2 June 1974.
39. *New York Times*, 7 December 1972.
40. *New York Times*, 30 January 1973.
41. *Newsweek*, 10 January 1972, p. 44.

industrialized nations of trying to stunt Third World progress and flauntingly invited companies having troubles at home with new antipollution laws to redirect their investments toward Brazil. In early 1972, the Brazilian minister of planning reduced the meager antipollution standards that did exist in order to promote foreign investors to proceed with the construction of wood-pulp plants along the coast.[42]

It may be, however, that the international issue-making dynamics fueled by United Nations meetings and communications fostered by international voluntary organizations and governmental agencies, coupled with pressure from Brazilian city governments plagued with pollution and from Brazilian scientists, all are having at least modest political effect. In March 1973, almost a year after its contemptuous stand on pollution havens at Stockholm, Brazilian military ministers began making public statements that suggested that no-holds-barred economic growth would no longer be the sole policy priority. The minister of the interior, an army colonel, told the São Paulo Engineering Institute that industrial development would have to be "adjusted to the preservation of the environment." [43] A month later the minister of industry and commerce met with leaders of a Japanese consortium planning to build a large cellulose factory in Brazil and informed them that their plans would have to include "the most advanced anti-pollution safeguards." The regime went on to generalize this policy, announcing that in the future all projects would have to gain the approval of the Industrial Development Council which would establish antipollution norms.

Such a marked change in direction will have to be accompanied by far-reaching political change, which may prove difficult for the military regime. Awareness of

42. Burns, "Brazil," pp. 18–19.
43. *New York Times*, 11 March 1973.

pollution effects and imposition of controls would mean curtailing the influence and perhaps the friendly access of domestic and foreign business interests in the circles of government decision making. It may also mean greater sensitivity at the center to public opinion at the state and city levels. And, finally, it requires new criteria for measuring the government's success; and if the quality of air and water are to be taken into account, how long can the maldistribution of wealth be ignored?

## CONCLUSION

There appears to be nothing inevitable about environmental deterioration accompanying Third World commitment to rapid economic growth. Singapore has adopted a government policy which has made it the refining and financial center of Southeast Asia, but has accompanied that policy with some of the most stringent environmental laws in the Third World. Overseas Chinese, who make up 85 percent of this independent island nation on the tip of the Malaysian peninsula, are often accused by their wary neighbors of being crassly materialistic and preoccupied with the pursuit of commerce. But the regime headed by Prime Minister Lee Kuan Yew has seen the integral connection between economic growth and pollution and has thus taken early steps to ensure that industrialization is matched by effective controls.[44]

44. See James Morgan, "Island of a Million Dustbins," *Far Eastern Economic Review*, 22 January 1972, pp. 26–27; see also *New York Times*, 21 June 1970 and 4 January 1971. For an analysis of how antipollution laws have been implemented in the Philippines, see Reynaldo M. Lesaca, "Pollution Control Legislation and Experience in a Developing Country: The Philippines," *Journal of Developing Areas* 8 (July 1974): 537–56. Also see Ross Marlay, "Politics and Administration of Environmental Legislation in the Philippines" (Paper presented at the Midwest Conference on Asian Studies, Chicago, November 1974).

Several conditions appear to have made the Singapore political system more responsive to environmental problems than has been the Brazilian political system. First, the regime of Lee Kuan Yew governs a small island whose natural resources are almost nil and whose limitations are far more blatant than in the territorially vast Brazil. Second, while Brazil includes several large urban centers from which the recent political pressure for pollution controls have come, Singapore is a city-state in which virtually the entire population is urban and thus especially vulnerable to pollution's effects. Both regimes have adopted development strategies that depend on foreign capitalist investment, but Singapore's regime has been less caught in any one great power's orbit and has thus cultivated national independence more energetically. Fourth, while both Lee Kuan Yew and the Brazilian generals have governed with the aid of authoritarian methods, Lee has utilized an elected party machine rather than an army.

Brazil and Singapore, along with Nigeria, Malaysia, Trinidad, and South Korea, are not typical of all developing nations. Their environmental dilemmas involve choices between rapid and controlled economic growth. Their choices and risks involve some of the same elements as environmental politics in affluent countries: man-made pollution, industrial laissez faire, overcrowded cities, citizens dependent on salaried jobs. But there are other developing countries whose governments are more worried about starvation than unemployment, more preoccupied with three-year droughts than with automobile congestion. These nations, found primarily among what the UN lists as the world's twenty-five poorest countries, should perhaps be treated separately when analyzing environmental politics. If Singapore or Nigeria represents the Third World, then Chad, Mali, and Nepal should perhaps be labeled the "Fourth World." In these countries environmental politics must be concerned chiefly with coping with natural hazards and devising man-made responses that do not exacerbate already extremely fragile conditions. In the Sahel, that belt of countries across Africa which has suffered lengthy droughts and yearly expansion of the Sahara Desert, environmental politics means not again repeating the mistake of experts who promoted goat herding among African nomads only to discover that during the droughts the animals, on which the herdsmen were now economically dependent, were drinking the little bit of water that would keep alive men and women.

In the Fourth World, environmental politics means improving the competency and coordination of international emergency assistance programs so that free food isn't sold for a profit and supplies left in warehouses to rot and rust because of the lack of effective distribution systems. In the Fourth World, environmental politics means that governments such as the Ethiopian will not try to hide the facts of its own drought and famine for the sake of maintaining its foreign tourist trade.[45]

Understanding the reluctance of developing country governments to treat man-made environmental disruptions seriously calls for a more rigorous and politically realistic comprehension of the dilemmas of development itself. The gaps between rich and poor in the world will have to be treated as seriously by industrial powers if they expect ex-colonies to enact controls that might retard their own economic growth. Just as hard, politicians and technocrats in both developing and developed countries will have to devise new definitions of measurements for "development"—ones that go beyond GNP to look at the distribution of growth benefits within a single

---

45. A detailed account of the ineffectual responses to the Sahel tragedy is included in Martin Walker's "Drought," *New York Times Magazine*, 9 June 1974, pp. 11–13, 42–46.

nation and the environmental costs entailed in economic expansion.[46]

46. A provocative effort in this direction is Edward J. Woodhouse's "Re-visioning the Future of the Third World: An Ecological Perspective on Development," *World Politics* 25, no. 1 (October 1972): 1–33.

# COUNTRY
# CASE STUDIES

PART **II**

The four chapters that follow are devoted to detailed analyses of environmental politics in the United States, USSR, Japan, and Britain.

In each case, two basic questions are asked. First, if you look at the fate of environmental issues in country X, what will you discover about that country's politics *in general?* Specifically, do environmental politics force us to revise the standard political portrait of country X? If you are an environmental advocate, will American politics look as pluralistic as presumed in textbooks; will Soviet politics be as closed and totalitarian as many observers would have us believe?

Second, what peculiarities of each system affect the outcome of environmental debates? Have there been differing political responses in each system to environmental issues? What pockets in each system have offered the most support or resistance to governmental attempts to preserve the environment against man-made hazards —local governments, political parties, courts, bureaucratic agencies?

Each of the four cases is organized around the topics introduced in earlier chapters—issue creation, group interactions in the policy process, bureaucratic politics, and development. This format should encourage comparisons among the four countries. More important, these subheadings permit lines of comparative questioning that can bridge the all-too-frequent gap between general analysis and specific cases.

Some questions that might be kept in mind as one moves from country to country include:

1. Has the emergence of environmental issues caused genuine change in the nation's politics, or is it simply another issue in "the same old game"?
2. When do political parties matter in defining and resolving environmental controversies?
3. Why do powerful bureaucratic departments in so many systems oppose environmental programs?
4. What political innovations in one country—e.g., the superdepartment in Britain, the "environmental impact statement" in the United States— might be most exportable to other nations?
5. Does it really matter for environmental protection whether commerce and industry are privately owned or run by the state?

# POLLUTION POLITICS IN THE UNITED STATES

.

## CHAPTER 5

## ENVIRONMENT IN A PLURALIST POLITY

American environmental politics has its roots in the conservation movement of the early twentieth century. Tied to the reformist Progressive movement's critique of capitalist "robber barons," fired by the end of the western frontier, and led by such gifted politicians as Gifford Pinchot and Theodore Roosevelt, the conservation movement was one of the strongest political forces in the nation. But fifty years later, in 1954, the Sierra Club, direct descendant of the conservationists, had a mere 8,000 members.[1]

---

1. Grant McConnell, "Prologue: Environment and the Quality of Life," in *Congress and the Environment*, ed. Richard A. Cooley and Geoffrey Wandesforde-Smith (Seattle: University of Washington Press, 1970), p. 4.

Issues, like nations, rise and fall; their careers rarely follow steady paths. The conservation issue had to do with natural resource exploration, opening of park land, and preservation of endangered species. It was essentially a nonurban movement, attracting western politicians and western interest groups. The arena for the conflicts were the U.S. Department of the Interior and the Interior Committees of the House and Senate, all dominated by westerners.[2]

Environmental issues arose in the United States out of a convergence of discrete events and a more general shift in society's values. By the 1960s the country was overwhelmingly urban and industrialized, dependent on the automobile and vast amounts of fossil fuels. Spreading public education and political mobilization were accompanied by a "power deflation" of traditional veto groups, all of which permitted greater access into political circles for new interest groups.[3] The environmental movement under these circumstances had a broader definition of issues than its conservationist predecessor. It posed questions more ideologically disquieting, questions regarding the distribution of power and wealth, the goals of productivity, and the necessity of economic growth for national well-being.

Legislation controlling environmental deterioration existed prior to the emergence of the movement, but it was piecemeal and frequently unenforced. The Rivers and Harbors Act of 1899 forbade the discharge of materials into waterways that would be hazardous to navigation; the Public Health Act of 1912 provided authority for federal investigation of water pollution that related to human diseases; the Oil Pollution Act of 1924 prohibited

---

2. Frank E. Smith, *The Politics of Conservation* (New York: Pantheon, 1966). See also Barbara Rosenkrantz and William Koelsch, eds., *American Habitat* (New York: Free Press, 1973).

3. Walter A. Rosenbaum, *The Politics of Environmental Concern* (New York: Praeger, 1973), p. 57.

oil discharges in coastal waters; the Federal Water Pollution Control Act of 1948 and its 1956 amendments gave government authority to impose sanctions against polluters, though it left prime responsibility to the states.[4] Between 1951 and 1962 only eleven states had any law relating to air pollution. The Donora, Pennsylvania, air pollution tragedy occurred in 1948, but not until 1955 did Congress pass effective air pollution legislation. In the mid-1950s, when the Sierra Club boasted its 8,000 members and pollution laws were fragmented and without teeth, the federal budget allotted only 3 percent of annual funds for environmental programs.[5]

The opening of the new era of environmental politics is impossible to pin down accurately. Some refer to the publication of Rachel Carson's controversial book *Silent Spring* in 1962, detailing the harmful effects of pesticides.[6] Others refer to the air pollution alert in Birmingham, Alabama, in 1971, which brought in federal authorities to shut down the city's giant steel mills. Still other commentators cite the Storm King controversy which became the rallying point for a political coalition that went well beyond the traditional conservation alliances. The Santa Barbara offshore oil blowout, which poured 235,000 gallons of crude oil on some of the nation's most beautiful beaches, also did its part to spark public and governmental attention.

What is clear is that by 1970 there was a new issue that was, in fact, a complex network of issues. In that year President Nixon delivered the first Message on the Environment to Congress, to be followed shortly thereafter by the nation's first Earth Day, organized by a

4. Philip P. Micklin, "Water Quality: A Question of Standards," in Cooley and Wandesforde-Smith, *Congress and the Environment*, pp. 132–33.

5. Rosenbaum, *Politics of Environmental Concern*, p. 11.

6. An analysis of the political aftermath of Rachel Carson's book is Frank Graham, *Since Silent Spring* (Greenwich, Conn.: Fawcett, 1970).

diverse coalition of interest groups. Efforts to pass comprehensive environmental legislation had already produced the National Environmental Policy Act of 1969 (NEPA). Title I of NEPA for the first time spelled out a national policy that acknowledged the impact of man's activities on the natural environment and made the federal government responsible for using "all practicable means consistent with other essential considerations of national policy" to ensure that "each generation of Americans will have a safe, healthful, productive and aesthetically and culturally pleasing" environment.[7]

Title II of NEPA established the Council on Environmental Quality, headed by a presidential appointee and having wide advisory authority. More important, NEPA required that all federal agencies or recipients of federal funds submit environmental impact statements detailing the probable effects on the environment of any forthcoming project. In 1970 the Congress created the Environmental Protection Agency (EPA) to monitor these crucial impact statements and to act as the federal government's chief enforcer of antipollution laws, such as the 1970 Clean Air Amendments and the 1972 Water Pollution Control Act Amendments. Only two years later, however, both EPA and the laws it was supposed to enforce were under attack from a wide spectrum of critics who claimed that the worldwide shortages of energy and other vital raw materials made environmental concerns secondary to the nation's top priority of sustaining economic growth.

The false starts, lags between environmental trage-

---

7. Quoted in Donald R. Kelley, Kenneth R. Stunkel, Richard R. Wescott, "The Politics of the Environment: The U.S.A., U.S.S.R., and Japan" (Paper presented at the International Political Science Association Meeting, Montreal, 1973), p. 4. This paper has been published in *American Behavioral Scientist* 17, no. 5 (May/June 1974). For a fuller treatment see Donald Kelley, Kenneth Stunkel, and Richard Wescott, *Economic Super Powers and the Environment* (San Francisco: W. H. Freeman, 1975).

dies and government response, and the waxing and waning of the issue in the public's mind suggest something not only about the fragility of environmental politics but also about the imperfections of the American political system.

The American political system is most commonly portrayed as a pluralist marketplace of ideas and interests. Environmental politics is often analyzed as one more example of pluralist politics at work.[8] To judge whether environmental politics in the United States does indeed fit this alleged pattern, it might be well to list briefly the major ingredients in a model pluralist democratic system:

1. The political decision-making process is open, with policy makers accessible and accountable.
2. On any given issue there will be a multiplicity of interests taking various sides in debate.
3. The various groups with a stake in an issue will have roughly equal resources with which to make their stands heard in policy circles.
4. People who lose on some policy questions will win on others; no one group will win all the time.
5. Most issues will be resolved through a process of *bargaining* and *compromise*.
6. Groups will constantly *shift alliances* as issues change.
7. Some of the conditions that promote and sustain pluralist politics will include fragmented power, decentralization, weak or ambiguous ideology, social mobility.

The American political system has its share of these conditions, perhaps more than most countries. Measured against this model, certainly American politics and Amer-

---

8. Rosenbaum, *Politics of Environmental Concern*; Donald R. Kelley et al., "Politics of the Environment," pp. 1–4.

ican environmental politics in particular is more pluralistic than those of the three other countries analyzed in the subsequent chapters.

Nevertheless, while pluralist relative to many other nations, U.S. politics fall far short of the ideal. To appreciate the context in which environmental decisions are made or thwarted, one has to clarify the distinctions between the ideal and the American pluralist reality. Reality looks more like this:

1. Many decisions are made by career bureaucrats who are only indirectly accountable to the public via the President and congressional committees.
2. Bureaucrats are accessible to some interest groups but rarely to all.
3. Congress works through specialized committees which often reach decisions behind closed doors and are cultivated by a select assortment of interest groups.
4. Presidents work increasingly through White House staffs which are not publicly accountable and which can override Senate-approved cabinet officers.
5. On many issues only one side has effective group representation.
6. Although there are thousands of interest groups in the United States, they are unequal in their financial, professional, or media resources.
7. The very institutional fragmentation and decentralization that provides so many access points for interests means that many groups are stretched beyond their capacities, unable to keep track of fifty state governments, various levels of courts, and the myriad of federal bodies that all may effect a matter such as environmental control.

8. In so complex a political system a citizen rarely has the information or spare time to follow his own interests effectively.

## CALIFORNIA: A LESS-THAN-IDEAL DEMOCRACY

One of the most open and pluralist states is California, which instituted many political reforms in the early twentieth century, making it the bane of political organizers used to working in the more closed systems of the East. The careers of two different environmental initiatives placed on the 1972 California ballot, however, underscore the imperfect character of American pluralist politics.[9]

Partly because of its rapid population growth after World War II and partly because of public awareness of and appreciation for the state's famed climate and scenery, environmental deterioration had become a more salient political issue in California in the early 1970s than it had in many parts of the United States—Los Angeles was better known for its smog than for its movie actors; the Sierra Club's active role in fighting for parks and against developers made it the best-known environmental advocate in the country; San Jose college students won national media attention when they ritually buried a new automobile; the 1969 Santa Barbara coastal disaster made "oil spill" a household word. Yet the California state government had done relatively little to translate this apparent public concern into concrete legislative proposals. In many less open states the courts and the

---

9. This case is based on Carl E. Lutrin and Allen K. Settle, "The Public and Ecology: The Role of Initiatives in California's Environmental Politics" (Paper presented at the annual meeting, American Political Science Association, New Orleans, 4–8 September 1973).

legislature would have remained the chief arenas for such translation. But California is one of twenty states that permit citizens to place proposed legislation on the ballot for consideration by the voters. Thus, even though state institutional sluggishness undercut pluralist expectations, there was an alternative avenue for translating opinion into policy and ensuring an energetic contest between interests.

In 1972 California voters faced two environmental referenda. In the June primary they were asked to vote yes or no on the "Clean Air Initiative" (Proposition 9), limiting the extraction of oil and gas from the state's coastal tidelands, imposing new restrictions on gas consumption and pesticides, and providing penal sanctions and civil penalties for all violators. Several months later, in the November election, Californians had to decide on the "Coastal Zone Conservation Act" (Proposition 20), which would establish a statewide commission plus six regional commissions to submit a land-use plan to the state legislature by 1976. This proposition would also prohibit any development within 300 feet of the coastline without a permit by either the state or regional commissions. A comparison of Proposition 9's defeat and Proposition 20's victory tells us something about the flawed character of American pluralist politics even in the most popular of forums, the referendum.

Primary elections traditionally elicit less public attention than do general elections. Consequently, California voters remained largely unaware of the meaning of the June Clean Air Initiative during most of the spring campaign. Only during the last month did the proposition gain sufficient visability to provoke close study by the citizens whose lives would be affected by its defeat or passage. The campaign that did attract the public's attention to Proposition 9 was fueled largely by an antiproposition alliance called Californians Against the Pollution Initiative, which included oil, chemical, and utility companies that would be most restrained by the proposed controls. During the week of May 5, at the height of the primary campaign, for instance, the People's Lobby which backed the proposition spent $3,000, while the opponents of Proposition 9 spent over $200,000.[10] The People's Lobby not only lacked the money needed to cover the large state with billboards, TV, and newspaper ads, it was also short on staff and on allies. The Sierra Club and the Audubon Society, for example, withheld their active support from the proposition, perhaps because of resentment against the newcomers moving into their issue domain. In the end the California electorate was especially influenced by the high-powered media campaign during the month just before the June primary, and this one-sided campaign effectively shifted the large mass of undecided or vaguely favorable voters into a majority against the Clean Air Proposition. As the respected director of the California poll, Mervin Field, concluded, "Many have a limited and even mistaken understanding of the issues. Opinions of this type are subject to quick change under the pressure of a massive campaign and emotional appeals." [11]

Circumstances putting the Coastal Zone Conservation Act, on the November 1972 ballot, were similar: increasing public awareness of one aspect of environmental deterioration plus an apparent state government stalemate. The Coastal Alliance, which secured the 403,000 signatures to get the initiative on the ballot, was dissat-

10. Ibid., p. 15.
11. Ibid., p. 16. A Japanese political scientist who analyzed the Proposition 9 campaign concluded that, though the environmentalists lost, the conflict itself served to keep the issue alive. See Akira Nakamura, "The Politics of Air Pollution Control in Los Angeles and Osake: A Comparative Urban Study" (Ph.D. dissertation, University of California, Los Angeles, 1973).

isfied with the continual bargaining and compromising on what it considered "already overnegotiated bills." [12] Like Proposition 9, coastal conservation proposal faced an electorate that was being asked to make up its mind not only on this sophisticated piece of legislation but on numerous other propositions (e.g., on the death penalty and on limitation of property taxes) as well as on a host of candidates for office. Likewise, early polls showed that there was low public awareness of the various propositions on the ballot, with those voters of higher income and education brackets expressing greater awareness. Five weeks before the November election, a "vast majority of the voters still had not made up their minds regarding five of the more controversial measures." [13] Thus, once again it seemed as though the stage were set for a media campaign, dominated by the interests with the greatest money and organization, which would be able to persuade a large proportion of undecided or confused voters.

But the Coastal Zone Conservation initiative did not suffer the same fate as the Clean Air Initiative five months earlier. The major difference was that the coastal proposition did not lose as many voters during the last month of campaigning. Analysts found that the major factors contributing to its success included: "(1) the large number of volunteer workers, (2) greater legal advice, (3) support from campaign firms, (4) positive wording of the proposition, (5) equal time on television networks to counter opposition arguments, and (6) unified efforts among ecology groups." [14] Proposition 20's opposition, composed of the California Builders Association, the California chambers of commerce, power companies,

---

12. Ibid., p. 21.
13. Ibid., p. 27.
14. Ibid., p. 31. For a description of the difficulties encountered in implementing The Coastal Zone Conservation Act during its first two years, see Rasa Gustaitis, "The Fight Over 'Improving' the California Coastline," *Washington Post*, 18 August 1974.

coast timber industries, local governments, and the County Supervisors Association and labor unions spent as much money in their campaign as they had to fight Proposition 9, almost $2 million. But this time the supporters of the initiative had more resources of their own to draw upon, as well as a more clearly defined proposition to communicate to the man on the street.

The California cases suggest that issues are not easily translated into authoritative decisions simply because lawmakers are elected and the public can make some policy directly.

### ISSUE RIVALRIES

For the environmental advocate, even the limited pluralism that exists in the United States has a dark side as well as a light side. Pluralism lends openness to a political system, permitting concerns to be translated into government policy proposals as the public develops new need or values. On the dark side of the pluralist "moon," however, is the multiplicity and fluidity of issues that threaten environmental issues being swamped by other questions crowding the political stage. The environmental issue's stiffest competitors have been crime, an issue not immediately connected to it, and energy shortages, an issue seemingly in direct conflict with environmental concerns. Despite competition, studies demonstrate that environmental matters have been making a steadily increasing claim on the American public's political attention.

Table 5.1 shows the rise in environmental issues between 1964, when pollution was mentioned as an important issue by none of Gallup's respondents, and 1971, when still in the midst of the Vietnam war, 7 percent mentioned pollution as a primary concern.

*Table 5.1.*  **Americans' Fears**

*Gallup Question:* What are your fears and worries about the future? In other words, if you imagine your future in the worst possible light, what would your life look like then? Again take your time in answering.

|  | *1959* | *1964* | *1971* |
|---|---|---|---|
| Ill health for self | 40% | 25% | 26% |
| Lower standard of living | 23 | 19 | 18 |
| War | 21 | 29 | 17 |
| Ill health for family | 25 | 27 | 16 |
| Unemployment | 10 | 14 | 13 |
| Economic instability | 1 | 3 | 11 |
| Unhappiness for children | 12 | 10 | 8 |
| Drug problem in family | — | — | 7 |
| Pollution | — | — | 7 |
| Political instability | 1 | 2 | 5 |
| Crime | — | — | 5 |
| No fears at all | 12 | 10 | 5 |

*Source:* Hazel Erskine, "The Polls: Hopes, Fears and Rights," *Public Opinion Quarterly* 37, no. 1 (Spring 1973): 140.

In 1971 *Newsweek* asked new young voters how they would like to see government tax revenues spent. Air and water pollution control received the largest backing, with 78 percent of the young people polled saying they favored more spending on pollution programs and only 2 percent saying they preferred less money spent.[15] When the spotlight shifted from age group to town size, pollution's issueness slipped. Residents of big cities were preoccupied with crime, though they did express more concern about pollution than did Americans living in rural areas, a fact underscored in New York City. In November 1973, at the start of Mayor Beame's term, New Yorkers were asked: "What are the 2 or 3 problems or issues of greatest concern to you personally that you would like to see the new city administration do something about?" In reply, 63

15. "New Priorities: How the Young Would Like to See Tax Money Spent," *Newsweek*, 25 October 1971, p. 44.

percent of the New Yorkers listed crime. Drugs and inflation followed in frequency. Much farther down the list was pollution, mentioned by only 9 percent—the same frequency rate as governmental corruption, but more than either welfare abuse or racism.[16]

When questions are asked of the American public concerning pollution alone, rather than pollution relative to other issues, there is evidence of growing concern and, more important in a politician's eyes, growing willingness to make sacrifices for the sake of protecting the environment. The Environmental Protection Agency, which naturally has a stake in demonstrating public concern in environmental programs, conducted a study entitled "The American People and Their Environment in 1973." [17] EPA surveyors questioned 3,012 adults in June 1973 about their views on pollution. The study revealed that more people claimed that their environment was getting worse rather than better. Furthermore, the more urban the area, the more likely respondents were to see deterioration.

Turning to the willingness-to-pay indicator, the EPA poll showed that, on the average, people volunteered to pay $62 per year for antipollution devices in a new car. To control air pollution at electric power generating plants, they were willing to bear an increase of 22 percent on their monthly bill. Finally, to reduce water pollution caused by food production and processing, the respondents agreed to pay $37.43 more per year. Not surprisingly, some sectors of the public showed greater willingness to pay than others. Households with higher incomes were

16. *New York Times*, 16 January 1973. A Gallup poll of large-city dwellers across the United States revealed as well that crime was the problem most mentioned (21%), followed by transportation and traffic (10%), juvenile delinquency (6%), high taxes (6%), community services (5%), and pollution, poor housing, and education (each 4%). "Problems based on Residents of large Cities," 1973, *Gallup Opinion Index*, Report No. 91, January 1973, p. 16.

17. Reported in "Intelligence Report," *Parade Magazine, Boston Globe*, 24 February 1974, p. 8.

willing to pay more for the sake of antipollution measures. Age was a second factor. People under 30 and those
between 55 and 64 volunteered to pay more than the
average.

All these polls were conducted before the American
public was confronted with the 1973–74 so-called energy
crisis. The Arab oil embargo combined with the even
more important underestimations of the escalating demands for petroleum created a new issue just at a time
when the American public and their representatives
seemed on the brink of investing attention and money in
environmental programs. Unlike crime, which frequently
(though not inevitably) distracted citizens' attention, energy shortages appeared to conflict with environmental
issues. Environmental groups found themselves blamed
for the long lines at gas stations because they were
fighting construction of new oil refineries. The Environmental Protection Agency felt the shift in issues so
acutely that it was compelled to extend deadlines for
clean air standards implementation. Congressmen un-
comfortable with the environmental movement now
could take the stump in the name of the Alaskan oil
pipeline with less fear of election-day retribution. The
auto industry and the public utilities could contend that
they were in favor of a clean and safe environment like all
Americans, but that in this mobile and industrial society,
transportation and development of nuclear reactors
would have to command top priority. In 1975 President
Ford could make a substantial delay in implementation of
auto emission standards part of his "energy package"
without fear of political backfire.

The energy issue's potency came from several factors: first, its impact was universal, affecting virtually all
sectors of this otherwise heterogeneous citizenry. Second, because of the high degree of automation and
economic integration marking the American society, en-

ergy shortages were experienced not through abstract charts or scientific health measurements, but on the most mundane and daily levels of individuals' lives. Third, the effects of energy shortages can be measured in terms of concrete things such as jobs and profits.

At bottom, the energy shortages and the deterioration of the natural environment do not *have* to be political rivals. Some environmental spokesmen warned that the energy shortages were one more sign that the traditional American ideology of growth was inadequate. But the *politics* of the energy issue were such that Americans were asked to *choose* between the environment and energy, between pleasant surroundings and economic security. Cast in that light, individuals could hardly resist the latter. Furthermore, the issue dynamics pitting environment versus energy served to alter power relationships in the American government, weakening those agencies with environmental responsibilities and their congressional and group allies, while adding new influence to already powerful legislative business and bureaucratic forces that had a stake in exploitation of natural resources and pursuit of accelerated economic growth.

Even stickier for environmentalists have been those policy questions challenging entrenched ideological assumptions. Nowhere have environmental advocates been more frustrated than in their proposals which appear to modify Americans' rights of private property. As the scope of environmental problems took on more and more of a national character and subissues became so tightly intertwined that it was impossible to deal with each separately, environmentalists called for long-range national land-use planning of the sort common in Britain. Because it touched a sensitive American ideological nerve, the land-use planning legislation took four years from drafting to congressional floor debate. Its ramifications were profound enough to draw a myriad of actors into the debate—the White House, Interior Department, Senate and House Interior Committees, Council on Environmental Quality, labor unions, farm groups, chambers of commerce, realtors, and of course environmental lobbies. Land-use planning was not a total stranger when the controversial bill reached the House of Representatives in 1974. Congress had passed the 1972 Coastal Zone Management Act providing federal grants to states to encourage shoreline planning. The Clean Air Act of 1970 and the water pollution legislation of 1972 had called upon the EPA to lay down land-use restrictions in certain specified situations. Senator Henry Jackson of Washington, a presidential contender and chairman of the Senate Interior Committee, had shepherded land-use planning bills through that chamber on two separate occasions. President Nixon in three messages to Congress had advocated land-use planning as a tool for effective environmental protection.

Yet by the time the land-use bill had reached the House floor in June 1974, it had become embroiled in ideological debate so heated that some original backers shied away and the specific elements of the modest bill were lost in a cloud of abstractions. Conservative groups mobilized to defend private property as the bedrock of American morality. The bill's sponsors, such as Congressman Morris Udall, tried to reassure colleagues that the legislation would merely provide federal funds for states that would study land-use planning feasibility; if passed, the sponsors argued, the bill would not expropriate a single acre of private property. Opponents saw the legislation, nevertheless, as the proverbial "opening wedge" which in time would lead to violation not only of sacred property rights but of states' rights. Furthermore, land-use planning would nourish the already mammoth governmental bureaucracy. Given the tenor of debate, White House support faded. In the final House vote members were bitterly divided. The land-use planning bill went down to a narrow defeat, 211–204.[18]

Land-use planning remains a goal for environmentalists since it appears to provide a means of coping with urban sprawl, overlapping jurisdictions, economic interdependency, and interlocking ecological conditions. To be translated into law, however, land-use planning proposals and their backers either will have to confront American ideological obstacles head-on, trying to alter basic public assumptions, or they will have to find a way of defining land-use planning that somehow reconciles it with established values. In the meantime, The Environmental Protection Agency is promoting land-use planning "through the back door" by withholding sewer construction funds to towns which are growing too fast.

The future of the environmental issue in American politics will depend, therefore, on a number of conditions:

1. Development of measures that make clear the costs of environmental deterioration so that they

---

18. For descriptions of the political jockeying that occurred during the congressional debate, see the reports of Gladwin Hill in the *New York Times*: 3 September 1973, 4 September 1973, 20 March 1974, 13 June 1974. A perceptive analysis of the pitfalls facing the bill's backers is Leonard Downie, "The Ambush of the Land Use Bill," *Washington Post*, 10 March 1974.

Studies of local and state land-use planning politics include David Jacobs, "The Long View of Long Island," *New York Times Magazine*, 17 February 1974, pp. 28–37; Spotlight Team, "Vineyard Judge Mixes Court Duties, Private Business," *Boston Globe*, 20 August 1973; Edward Walsh, "Logrolling for Land Use Bill [in Maryland]," *Washington Post*, 24 March 1974; Calvin Trillin, "U.S. Journal: The Coastline [Martha's Vineyard]," *New Yorker*, 18 November 1972, pp. 215–24; Jon Nordheimer, "Florida Seeks to Curb Runaway Growth," *New York Times*, 19 November 1973; Robert C. Fellmeth, *Politics of Land: Ralph Nader's Study Group Report on Land Use in California* (New York: Grossman, 1973); William Osborne, *The Paper Plantation: Ralph Nader's Study Group Report on Land Use in Maine* (New York: Grossman, 1974). Shelley M. Mark, "Hawaii Land Use Planning and Control," *HUD Challenge* 4, no. 10 (October 1973): 7–11.

seem as immediate as the costs of street crime.
2. Access of environmentalists to media, money,
   and staff with which to communicate these
   findings to the general public and respond to
   opponents' challenges.
3. In the short run, lessening of popular anxieties
   over crime and energy so that attention can be
   devoted to water and air pollution and misused
   land.
4. Or: public integration of distracting or competing
   issues into a broad coherent set of political
   demands.
5. Restatement of American ideology in a fashion
   that does not equate public goals with unrestrained private enterprise.

## ISSUES AND FEDERALISM

It has been to environmentalists' advantage to have
what amounts to fifty separate political laboratories in
which to experiment with various lobbying or enforcement strategies.[19] If the Clean Air Initiative loses in
California, then clean air lobbyists elsewhere will at least
have lessons to draw upon. If a law prohibiting no-return
bottles proves effective in Oregon, then activists can
wave it in front of doubting Thomases in New York and
Vermont.[20]

---

19. For a valuable survey and comparison of various states'
approaches to environmental control, see Elizabeth H. Haskell and
Victoria S. Price, *State Environmental Management: Case Studies of
Nine States* (New York: Praeger, 1973). For a more general comparative
analysis of state government efficacy and innovativeness, see Virginia
Gray, "Innovation in the States: A Diffusion Study," *American Political
Science Review* 67, no. 4 (December 1973): 1174–85. Gray concludes
that New York and California have been the most innovative, not
because they are the largest or wealthiest, but because they have the
most competitive party politics.
20. The Oregon "bottle law" passed in October 1972 prohibits
no-return bottles and cans and requires payment for empty containers

Furthermore, federalism permits environmentalists to define issues in ways that fit the peculiar circumstances of a relatively narrow electorate. Thus, for instance, the importance of environmental control was brought home to Coloradans by focusing on the damage to the state likely to result if the Winter Olympics were invited. Whereas in Delaware environmental issues were made salient by concentrating on the harmful effects resulting from siting nuclear reactors along the state's scenic shore.

The disadvantages of federalist fragmentation abound, however. The energy crisis of 1974 highlighted the ways in which federalism can split the public regionally. As public anxiety grew over scarcities, those states that were prime producers or refiners of fuels became increasingly annoyed at states which consumed great quantities of energy but refused to have "dirty" installations within their own borders. Louisianans wondered out loud why they should ship their oil and natural gas out of state and suffer shortages and high prices at home simply because New Englanders didn't want refineries or drilling rigs along their coast. Southerners were angered by federal regulatory orders that required southern gas producers to ship more gas out of state. Bumper stickers began appearing on cars in the South: *Let the Bastards Freeze in the Dark.* At the local level ironies were acute. Oklahoma farmers with oil wells on their own land were short of diesel fuel for their tractors and propane gas for their home heaters. Louisiana and Texas, which courted industry with promises of endless supplies of cheap natural gas, could not meet company demands. Louisiana's Governor Edwin W. Edwards, spokesman for the

---

returned to stores. The bill was passed over concerted opposition of beverage and bottling companies as well as grocery retailers. For descriptions of the political successes of opponents in preventing such bottle laws in other states, see *Washington Post*, 21 October 1973; *New York Times*, 8 July 1974.

region during the crisis, echoed the sentiments of bumper-sticker wearers when he explained, "We're not trying to be Arab sheiks down here, we just expect other states to do their share." [21]

In the context of federalism, the collision of energy and environmental issues spurred local chauvinism. As the editors of one respected journal asked their readers, "Shall we strip-mine Iowa and Illinois to air-condition New York?" [22] In Oregon and Colorado campaigns were mobilized to halt the influx of out-of-state newcomers. *Don't Californicate Oregon* signs made it clear that local residents had seen the possible future across their borders and were determined to avoid it even at the risk of jeopardizing federal cohesion. The U.S. federal system relies on the free flow of persons and trade. States eager to slow growth face formidable legal obstacles. But environmental consciousness is raising again questions as old as the nation itself.[23]

## ISSUES AND UNEVEN DEVELOPMENT

Maldistribution of jobs, leisure, education, and capital means that environmental issueness will be grasped in some sectors of the American public while shrugged off in others. Blacks, marginal workers—especially in company-dominated towns—and those residents of states just beginning to taste the rewards of industrialization have joined with corporate managers in voicing suspicions of the environmental movement. As long as Ameri-

21. Quoted in James P. Sterba, "U.S. Energy Crisis Stirs Self-Interest of Regions," *New York Times*, 20 December 1973.

22. Editors, *Atlantic*, September 1973, p. 84.

23. Christopher S. Wren, "I've Got Mine, Jack: Environmentalism, Colorado Style," *New York Times Magazine*, 11 March 1973, pp. 34–46. E. J. Kahn, "Letter from Oregon," *New Yorker*, 25 February 1974, pp. 88–99.

can pluralism coexists with economic and social inequality, the saliency of the environmental issue is likely to be spread unevenly.

In the early 1970s, as the South was experiencing its long-awaited economic and population boom, progressive southern governors began to take steps to avoid the environmental consequences that industry had brought to other parts of the country. In a sense, Governor Dale Bumpers of Arkansas and Governor Jimmy Carter of Georgia were in positions analogous to the generals and politicians heading governments in Brazil, Indonesia, and Mexico. They had to find an answer to the question that plagues all leaders whose citizens are on the verge of long-denied economic growth: How great a social and physical price would be paid for new jobs, capital, and security?

Governor Bumpers tried to reconcile environmental and economic needs by urging his Arkansas constituents to measure the worth of economic advances by their own rural standards. "I've admonished the people of Arkansas about boosterism for boosterism's sake." The governor continued, "We're no longer defensive about being a rural state. We are proud that we are a rural state." [24]

In the North the question of development's price has been raised most graphically in Maine. The state's unemployment rate was seventh highest in the nation in mid-1974, while its per capita income ranked the state forty-fifth out of fifty. Moreover, the sounds of state chauvinism heard during the winter energy crisis made northeastern states such as Maine nervous about the availability of fuel in the future. But if Maine was the equivalent of Brazil, several of the state's small coastal towns were the counterparts of Indian reserves in the Amazonian interior. For when coal and petroleum compa-

---

24. *New York Times*, 18 November 1973.

nies such as Gibbs Oil and the Pittston Corporation came to Maine to shop for new refinery sites, they came to such declining communities as Eastport and Sanford. Townsmen there had more to say about the acceptance or rejection of refineries than did Brazil's Indians confronting international timber companies and government road builders, but their dilemma was not dissimilar. For Eastport the decline of the fishing industry meant that the townspeople weighing the company offers were even more economically depressed than the state as a whole and their resources for controlling powerful outside developers even less. The debate in Eastport over Pittston's refinery split old-timers and newcomers, the former anxious to halt the town's decline by infusing new jobs and capital, whereas the latter were attracted to the area in the first place because of its removal from industrialization.[25]

Thus, the politics of legitimizing the environmental issue necessitates several things. First, environmentalists in Eastport and similar places have to avoid polarization between old-timers and locals versus newcomers and outsiders. Outsider-insider polarization can be diluted, for instance, by demonstrating that environmental activists are not the only non-locals, that companies such as Pittston are also based elsewhere and care chiefly about exploiting local resources for outsiders' profit. It can also be avoided by creation of ad hoc antirefinery coalitions within Maine itself. This may require keeping external agencies at arms' length even when, as in the case of the federal EPA, it is a potent ally, and instead bolstering Maine's own state environmental agencies. Here, too, the insider-outsider dichotomy can be diluted by pointing to

25. For a description of the local, state, and federal factors interacting in Eastport, Maine's, debate, see Tom O'Brien, "Oil Refinery Plan Revives a Dream for Maine Port," *Washington Post*, 19 May 1974. Also, John McDonald, "Oil and the Environment: The View from Maine," *Fortune*, April 1971, pp. 84–89, 146–50.

the outsider priorities of prorefinery allies such as the Federal Energy Office. A second strategy for legitimizing environmental concerns in economically depressed areas is to demonstrate the limited economic benefits that will actually come into the area if the environmental risks are taken. Will Pittston, for example, create jobs at its refinery but then hire out-of-state skilled labor for those jobs? Will Pittston be taxed in such a manner that its coming to the region will ensure improved public services?

The United States is economically divided socially as well as geographically. Ethnic or class groups that are economically depressed see environmentalism as an obstacle to new jobs and a threat to existing jobs. In Jacksonville, Florida, for instance, the recently cultivated biracial politics has taken the form of brushing aside environmentalist warnings for the sake of luring new industry to the city. Jacksonville's black leaders joined with its white commercial and political elite to back a joint Westinghouse-Tenneco factory for assembling floating nuclear power plants. The Florida Audubon Society filed suit against the corporations and the Army Corps of Engineers, responsible for clearing an environmental impact study, charging that construction of the factory on 1,300 acres of filled land would endanger the richly productive Black River Marsh. But local black groups were more interested in job guarantees than in fish preservation and thus concentrated their energies on securing from Westinghouse-Tenneco the pledge that 23 percent of the factory's jobs would go to minority workers. Jacksonville blacks were aware that in adopting these priorities they were jeopardizing a valuable alliance. As the black coalition contended in its own court brief, however, the factory would have an impact not just on marsh ecology but on human ecology. One of the black lawyers explained, "We pondered a long time before we even considered joining the defense in this suit. . . . We didn't feel we could afford to lose our friends who are environmentalists. Some of these people are the very ones who have helped us pass civil rights laws and helped us walk the streets. But we decided the issue is of such momentous importance, will be such a tremendous step toward getting jobs, that we had to do it." [26]

Blacks, on the other hand, do perceive the relevance of environmental issues when they are directly related to urban safety and economic opportunity. Thus the campaign to enforce laws prohibiting lead-based paints in apartments has served as a slender bridge between black power and the environmental movement. Similarly, mass transit, critical for expanding job opportunities, has linked the two and has been a major issue for black politicians in cities such as Los Angeles and Atlanta.[27]

## GROUPS AND "SUBGOVERNMENTS"

In analyzing the emergence or blockage of political issues, we noted the imperfections in the American pluralist model: groups are unequally equipped, issues frequently are uncontested, the average citizen has too little time or skill to weigh complex questions that affect his own life. In the environmental policy-making process one discovers further shortcomings due to the existence of informal but nonetheless politically powerful "com-

---

26. Daniel Zwerling, "Florida Fight: Ecology vs. Jobs," *Washington Post*, 20 January 1974. For a similar definition of blacks' human ecological priorities, see Nathan Hare, "Black Ecology," *Black Scholar*, April 1970, pp. 2–8. Regarding the "ecology vs. jobs" debate in Gary, Indiana, see chap 1, note 36.

27. When initially proposed in a popular referendum, Atlanta's rapid-transit plan was defeated largely by blacks, who viewed it as a giveaway to white suburbanites. But in 1973 the referendum was revised with black consultation and passed in a city-wide election with overwhelming black support. Joseph Kraft, "Power to the People?," *Washington Post*, 4 February 1973.

plexes" that give certain groups more influence than others.

A less dramatic term than "complex" may be "sub-government," a term originally applied by Douglass Cater to describe networks of key actors that determine America's sugar-import quotas.[28] Cater discovered that a close interlocking of specialized congressional committees (in the sugar policy area, for instance the congressional Agriculture Committees), middle-level executive branch bureaus, and powerful commercial interest groups together hammered out U.S. policy, which the rest of Congress as well as the President simply rubber-stamped with only minimal active involvement.[29] These networks, or "subgovernments," exist in such diverse policy realms as Indian affairs, highway construction, industrial worker safety, and weapons procurement.

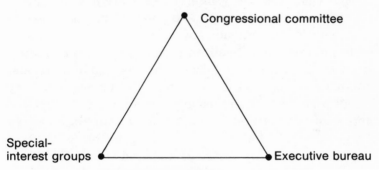

*Figure 5.1.* **A Subgovernment**

A subgovernment is likely to evolve under the following conditions:

28. Douglass Cater, *Power in Washington* (New York: Vintage Books, 1965).

29. In June 1974 the Congress took a step toward breaking up this long-standing subgovernment. The House as a whole overrode its Agriculture Committee and voted to terminate the Sugar Act which provided for a guaranteed market for domestic sugar producers and specific quotas for various foreign sugar-importing countries.

1. a relatively narrow policy field
2. specialized congressional committees responsible for that field and deferred to by the rest of Congress
3. unequally equipped interest groups in the field plus general apathy on the subject in the public
4. relatively autonomous bureaucratic agencies able to cultivate ties of their own outside the Executive

Where subgovernments grow up, they exacerbate existing inequalities of access and information in American politics. Subgovernments are valuable to their lobbyist, congressional, and bureaucratic participants because they keep potentially controversial or embarrassing issues "just among us boys," out of view of the press, the public, and even the White House. Carefully cultivated relations of mutual trust and interdependency among groups inside and outside government which have immediate interests in specialized policy areas help to reduce friction for the policy makers, act as a buffer against annoying outside intrusions, and build up predictability for all concerned. There usually are differences of opinion and even competition among actors within a specialized subgovernment, but it occurs among people who work together continually and have a shared stake in preventing wider participation in that field.

The interest groups most capable of cultivating such intimate relationships with Senate and House committees and their bureau counterparts are those possessing at least several of these attributes:

1. a clearly defined stake in the specialized area
2. legitimacy in the eyes of congressional committee members
3. money enough to afford offices in Washington

and staffs able to conduct research of use to the committee's work
4. additional funds with which to support individual committee members' electoral campaigns
5. organizational bases (clubs, business firms, etc.) at the local level from which the committee members themselves come

No committee-group-bureau network is unmovable. New associations form and seek entrée; congressmen die or are defeated; the President, party leaders, or press turn their attentions to policy areas that previously had been ignored by all but a few specialists; established interest groups shift tactics to meet new challenges. For example, the American Petroleum Institute, the spokesman for the major U.S. oil companies, reacted to the upsurge of public, White House, and politician interest in the effects of oil on the environment and economy by altering its lobbying techniques. With the increased political attention being given to oil policy questions, the API no longer could depend so heavily on its low-profile contacts with committees and middle-level bureaucrats. The API hired former Texas Congressman Frank Ikard as its new head, assuming that Ikard could give the API easier access into congressional offices, though some oil executives argued that Ikard has been out of Congress too long, that his valuable contacts as a member of the "club" are dwindling.

Ikard responded to mounting public criticism of the industry by moving API's offices from New York to Washington and by stepping up its lobbying activities. He spent more time on Capitol Hill in person and appeared regularly on television and before press conferences. During the energy crisis API added more lobbyists and increased its lobbying budget to $200,000 (out of a total 1974 association budget of $15.7 million), five times what it was two years earlier. Recognizing the inadequacy of staffing in Congress, the API built a policy-analysis division of its own which could supply technical aid to congressmen on specific bills. Finally, in reaction to the growing part played by the executive branch in energy policies, especially with the 1974 creation of the Federal Energy Office, the API directed more lobbying effort at the executive branch and recruited as its second-in-command Charles J. Di Bong, former consultant to President Nixon on energy affairs.[30]

Because environmentalists are typically excluded from pertinent subgovernments, they have had a special interest in altering subgovernment relations so as to open them to wider publics. Environmentalists have cultivated (but also publicly criticized) Senator Edmund Muskie, the Maine Democrat, because of his powerful position as chairman of the Senate Public Works subcommittee on Water Pollution. They have built up formidable research facilities in order to supply technical assistance to friendly committee members reviewing environmental legislation. They have donated money and volunteers to sympathetic encumbents on strategic committees and worked on campaigns of candidates challenging powerful committee members who have opposed stiffer environmental legislation. The environmentalists' most significant victory to date was their 1972 primary upset of Colorado Congressman Wayne Aspinall, chairman of the House Interior Committee and regular environmental opponent.[31]

30. *Washington Post*, 24 March 1974; *New York Times*, 4 February 1974.

31. In 1972 a coalition called Environmental Action distributed a list of encumbent congressmen labeled "The Dirty Dozen." Eight of the 12 managed to win reelection despite environmentalists' campaigns. Gladwin Hill, "Environment Vote a Factor in Fifty Congress Contests," *New York Times*, 12 November 1972. On the opposite side of the ledger, those congressmen found to have voted most consistently in favor of stiffer environmental regulation have been ideological liberals in both

Nevertheless, subgovernments evolve and block environmental legislation efforts not simply as a result of a handful of congressmen who place commercial exploitation over protection of nature, but because the Congress and the U.S. political system in general place great value on specialization and on achievement of concrete benefits for geographic constituencies. Analysis over several sessions of Congress reveals that committee memberships are distributed as the result of interaction of (1) the individual representative's own strategic evaluations of what will further his political career, and (2) the Republican and Democratic leadership's estimation of what committee assignments will enhance the party's and the leaders' own political leverage. Since it is western constituencies, for instance, that traditionally have had the most to gain from bureaus overseen by the Interior Committees—parks, public land leases, reclamation projects—westerners have long dominated the Senate and House Interior Committees where so many of the important environmental bills are considered.[32] On the other hand, in the powerful House Appropriations Subcommittee on the Interior, there has been some deliberate effort to limit the number of westerners, with the result that budgets for agencies such as the western-favored Bureau of Reclamation get approved in the House Interior Committee only to be cut back in the House Appropriations Subcommittee.[33]

---

parties but liberal Democrats in particular. Leonard G. Ritt and John M. Ostheimer, "Congressional Voting and Ecological Issues," *Environmental Affairs* 3, no. 3 (1974): 459–72.

32. Richard F. Fenno, *Congressmen in Committees* (Boston: Little, Brown, 1973), pp. 34–43; David W. Rhode and Kenneth A. Shepsle, "Democratic Committee Assignments in the House of Representatives," *American Political Science Review* 67, no. 3 (September 1973): 889–905.

33. Ira Sharkansky, *The Politics of Taxing and Spending* (New York: Bobbs-Merrill, 1969), p. 71. Between 1947–62 the Bureau of Reclamation's budgetary average growth rate was only 3.6 percent versus 23 percent for HEW's Office of Education and 17 percent for Interior's Bureau of Land Management. Ibid., p. 69.

One western member of the House Interior Committee gave his reason for selecting that assignment:

> I was attracted to it, very frankly, because it's a bread and butter committee for my state. I guess about the only thing about it that is not of great interest to my state is insular affairs. I was able to get two or three bills of great importance to my state through last year. I had vested interests I wanted to protect, to be frank.[34]

The preferences of westerners for the Interior assignment, plus the relative disinterest of eastern and urban representatives, made it easier for certain lobbyists to cultivate friendly relations there. An official of the American Mining Congress explained:

> In most instances, the Interior Committee is more sensitive to the problems of the [commercial] users than is the Interior Department. That's because they have to get re-elected by the citizens every two years. They get a better grasp of public attitudes than the people in the agencies, [who] . . . don't know what a ballot box or precinct is.[35]

The third link in a subgovernment is the specialized executive bureau, usually two or three steps below the presidentially appointed cabinet secretary. Contrary to constitutional mythology, American executive and legislative units do not view each other at dagger points. Concentration of a narrow field of policy creates bonds of respect and dependency. An agency such as the National Park Service must not only win authorization of its programs from the House and Senate Interior Committees, but also from the two appropriations subcommittees

---

34. Rhode and Shepsle, "Democratic Committee Assignments," p. 894.

35. Fenno, *Congressmen in Committees*, p. 42.

on Interior. All that is authorized is not appropriated. Thus an agency needs friends along the precarious budgetary route. Committee and interest-group support can also be valuable if the White House or department secretary decides to cut back the agency's functions or personnel.

One of the best documented environmentally related subgovernments is in the field of federal forestry policy. Figure 5.2 suggests the pattern of relationships that binds certain actors together and produces national forestry policy. The *dotted* lines refer to actors with a stake in forestry policy but who are excluded from the intimate network except at times of unusual public attention. Over the years, this subgovernment has supported a federal policy toward the 186 million acres of national forests that favors commercial leasing and exploitation and resists efforts to impost stricter conservation rules. Those groups outside the subgovernment are seeking more influence in order to compel the Forest Service to redefine its mission and to make the congressional agriculture committees which oversee the Forest Service less prone to accept the rationales and statistics of timber companies.

## PRESIDENTS AND SUBGOVERNMENTS

One orthodox remedy for subgovernment exclusiveness has been assertion of presidential authority down through the ranks of the bureaucracy. To break up or at least loosen the "forestry subgovernment," for example, a President might make natural resource conservation a publically stated White House priority and instruct White House aides and the secretary of agriculture to keep close check on the Forest Service so that it articulated that

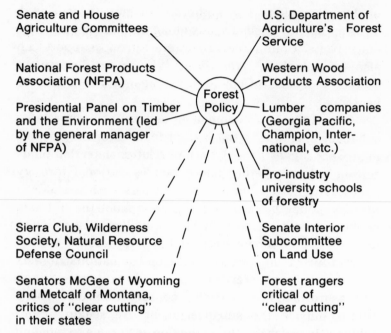

*Figure 5.2.*   **The Forestry Subgovernment**

Source: This diagram is based on Harmon Henkin, *The Forest for the Trees* (New York: Ballantine Books, 1970); James Risser, "The U.S. Forest Service," *Washington Monthly,* December 1971, pp. 16–26; James Risser, "Wasting the Wilds," *Washington Post,* 4 April 1971; Loran Stewart, "New Forest Law Needed," *New York Times,* 31 January 1971; Bartle Bull, "Smokey the Bear Sells Out," *Village Voice,* 17 August 1972.

stand in its dealings with Congress and lumber companies.

The Watergate scandals tempered some of the traditional enthusiasm for this "presidential solution," insofar as it alerted the public to the risks involved when civil servants are at the bid and call of White House politicians. But environmental politics, too, reveals flaws in this presidential remedy. The policies that subgovernments traditionally favor may be precisely those that the White House itself prefers, and thus there is little incentive for a President to intervene. In fact, the President may act so as to *reinforce* subgovernment relationships. This was the case when President Nixon appointed the general manager of the National Forest Products Association as the chief contact point between the White House and the Presidential Panel on Timber and the Environment. In addition, if a President elects one or two policy fields for his personal consideration—e.g., civil rights, foreign policy—he may prefer to let the rest of the bureaucracy operate on a relatively loose rein. During such periods of White House "benign neglect," civil servants and presidential executive appointees still have budgets to justify and clients in the field to mollify. Lacking clear signals from the White House, they move to protect their own agencies or departments by developing outside support.

The principal vehicle for ensuring some degree of White House coordination and discipline in the sprawling federal bureaucracy is the Office of Management and Budget (OMB), formerly the Bureau of the Budget. The OMB has become the central monitor of all bureaucratic programs, calling on agencies to list their specific goals, show how they match presidential objectives, and demonstrate periodic progress toward fulfilling those goals. Those who imagine that accountants are about as potent as Bob Cratchit on his Dickensian high stool sorely underestimate the real power of budgetary officials. In environmental affairs the impact of OMB is chiefly in the form of compelling a particular form of justification for programs—a justification that spells out costs and benefits of programs in dollars-and-cents terms and that puts greatest stress on short-term consequences.[36] Because

36. Allen Schick, "The Budget Bureau That Was," *Law and Contemporary Problems* 35, no. 3 (Summer 1970): 519–39; John Herbers, "The Other Presidency," *New York Times Magazine*, 3 March 1974, pp. 16–17, 30–41. The OMB's role in opposing EPA-backed Safe

environmental costs are often difficult to translate into
dollars and cents and because the harmful effects of a
dam or canal may not be apparent for a generation,
agencies tend either to neglect or understate environmental consequence of their projects in reports to the OMB.[37]

## BUREAUCRATS AND THE ENVIRONMENT

Ghosts were all around him. Washington, Jefferson,
Lincoln, Wilson, FDR, Kennedy and the great Sena-
tors, Webster, Calhoun, Clay, Norris, Taft. Of course
there were no monuments to civil servants, a fact in
which Weller took a perverse pride. The civil service
worked best when it worked in the shadows, truly a
government-within-a-government.[38]

American bureaucrats—or "civil servants" as they
prefer to be called—are vulnerable. Their best protective
device is anonymity. The second best is cultivation of
mutually beneficial bonds with those outsiders who care
most about the agency's actions and who control rewards
and sanctions relevant to the agency. The long-term
relationship that the Forest Service has built up with the
Agriculture Committee and the timber industry is not for
the sake of soliciting bribes or soft jobs after retirement
so much as it is for the sake of warding off disturbing
interference which destroys anonymity and disrupts orderly routines.
    The American bureaucrat feels vulnerable because, in

Drinking Water Act of 1974 is detailed in "Bill for Pure Water and
Standards Faces New Hurdles in Congress," *New York Times*, 6 October
1974.
    37. For a critique of the cost-benefit analysis in the Bureau of
Reclamation, see Richard L. Berkman and W. Kip Viscusi, *Damming the
West: Ralph Nader's Study Group Report on the Bureau of Reclamation*
(New York: Grossman, 1973).
    38. From Ward Just, "Weller in Love," *Washington Post/Poto-
mac*, 9 June 1974, p. 12.

a culture based on populist ideals, a nonelected official makes the public uneasy if not actually resentful. Moreover, the American bureaucrat does not enjoy the social prestige conferred on his counterparts in many other nations. He lacks the elite status of the British civil servant and the close ties with the ruling party possessed by Soviet and Japanese bureaucrats. Consequently, it is not surprising that federal bureaucrats whose careers have been with such agencies as the Forest Service, Bureau of Mines, or Army Corps of Engineers should view the environmental movement with wariness. This newly mobilized sector threatens to disrupt traditional patterns of communication and support, interjecting new pressures and demands, shedding the lights of the media on ordinary routines.

On the other hand, some bureaucrats may welcome the environmental movement's intrusion. These are officials who have suffered from insufficient support for their programs—e.g., the National Park Service—or who are working within agencies they believe are betraying this original mandate—e.g., forest rangers within the Forest Service, Indian rights advocates within the Bureau of Indian Affaires. Other bureaucrats who look favorably on the environmental movement are those in newly created agencies mandated to implement the NEPA but lacking established support bases.

Among the numerous agencies that handle environmental policy there are differences in institutional security, clienteles, ideology, budgetary resources, White House backing, and legal mandates. The principal agencies include:

### Environmental Protection Agency

Created in 1970 to administer the National Environmental Protection Act (NEPA) and the Water Pollution and Air Pollution Control Acts. Intended to combine several scattered agencies and to monitor compliance with federal law and to advise state governments on environmental programs. Located in the Executive Office and not of cabinet rank.[39]

### Council on Environmental Quality

Established in 1969 under the NEPA. Its three presidential appointees and a small staff screen governmental programs from the standpoint of environmental policy.

### National Park Service

Agency within the Interior Department. The service has been criticized by environmentalists for "giving away" park lands, currying favor with park concessionaires, and permitting parks to be "destroyed through piece-meal development." [40] These criticisms were directed against the career civil servant who last headed the service. Still, anxiety was expressed when the careerist tradition was broken by the Nixon appointment of a young White House aide to direct the service in 1972.

### Bureau of Sports Fishing and Wildlife

Founded by Theodore Roosevelt as part of the Interior Department in 1903. Now administers some 30 million acres in the wildlife refuge system. Criticized by Ralph Nader's investigators for laxness in estimating fish and wildlife costs of reclamation projects.[41]

### Bureau of Land Management

Agency of the Interior Department administering 450 million acres of public domain spread across the western

39. See James R. Michael with Ruth Fort, eds., *Working on the System, Ralph Nader's Center for the Study of Responsible Law* (New York: Basic Books, 1974), pp. 825–62.

40. Friends of the Earth spokesman quoted in *New York Times*, 14 December 1972.

41. Berkman and Viscusi, *Damming the West*, p. 59.

states and Alaska, greater in area than all the national parks and national forests combined. The agency is an advocate of "multiple use," which at times had been a euphemism for environmental disruption for the sake of mineral or fuel exploration through sales and lease of federal lands.[42]

### Bureau of Outdoor Recreation

Introduced into the Interior Department in 1962, its mission is to develop and maintain a nationwide outdoor recreation plan and to assist other agencies in dealing with national recreational resources. Congress gave the BOR responsibility to develop a master plan but failed to give it power to make other federal agencies, many of which resent its intrusion, follow the plan. BOR does, however, have the power to allocate funds to state and local agencies for projects conforming to the national plan.[43]

### Bureau of Reclamation

Created within the Interior Department under the Federal Reclamation Act of 1902 to promote economic settlement of the West, the bureau has been the chief builder of dams. With its formidable agricultural allies in the West, it has been charged with ignoring the environmental and larger social impacts of its irrigation and flood-control projects. The more celebrated of its controversies was with the Sierra Club over the damming of the canyons along the Colorado River.[44]

---

42. Fellmeth, *Politics of Land*, p. 259.
43. Ibid., p. 254. Budgetary politics involving the BOR are analyzed in: Jeanne Nienaber and Aaron Wildavsky, *The Budgeting and Evaluation of Federal Recreation Programs* (New York: Basic Books, 1973).
44. Berkman and Viscusi, *Damming the West*; John McPhee, *Encounter with the Archdruid* (New York: Ballantine Books, 1971), pp. 135–215; Frederick N. Cleveland, "Administrative Decentralization in the U.S. Bureau of Reclamation," *Public Administration Review* 13 (Winter 1953): 17–29.

### U.S. Geological Survey

The arm of the Interior Department responsible for inspecting and regulating oil and gas operations on the continental shelf and other public-domain lands. After the exposure of the custom by the House Subcommittee on Conservation and Natural Resources, the Geological Survey agreed to discontinue its practice of consulting in secret a committee of oil industry officials before establishing offshore drilling regulations.[45]

### Forest Service

The Department of Agriculture agency that oversees a national forest system covering 186 million acres in 40 states. Like the Bureau of Land Management, the Forest Service credo has been "multiple use." Environmentalists have proposed that the Service be transferred from the production-oriented Agriculture Department to the more conservationist Interior Department.

### U.S. Soil Conservation Service

A branch of the Department of Agriculture which has been charged with being "so intent upon harnessing water to serve big land holders' needs that it ignores the ecological side effects of its work." [46] Like the Bureau of Reclamation, the SCS has been charged with making a mockery of the White House-backed cost-benefit accounting system, inflating benefits, and excluding ecological costs.

### U.S. Coast Guard

Like many federal agencies, the Coast Guard's self-image and mission have been reshaped by environmental laws. The Water Pollution Control Act has widened its

45. *New York Times*, 26 April 1974.
46. Tom Herman, "Waterway Wrangle," *Wall Street Journal*, 19 July 1971.

law-enforcement responsibilities and it has become the key agency in the prevention and investigation of oil spills.[47]

### Army Corps of Engineers

Under the Defense Department but possessing autonomous clienteles and congressional ties, the Corps is among the most powerful practitioners of bureaucratic porkbarrel politics. It has reached a modus vivendi with the Bureau of Reclamation whereby the latter takes most projects in the West while the Corps concentrates on public works in the rest of the country. Reacting to unfavorable publicity, the Corps sought to redefine at least its image, stressing its role as the giver (or withholder) of permits to companies or towns discharging refuse into navigable waters, a role that the Corps had allowed to lay dormant.[48]

### Tennessee Valley Authority

An independent agency which operates the New Deal-inspired hydroelectric and flood-control project in the Tennessee Valley region. Over the last four decades it has become increasingly like the private utility companies, resisting environmental controls that would limit its output and revenues. It has also become one of the nation's major contractors for strip-mined coal. In early 1975, when it was announced the TVA's directors were considering buying the Peabody Coal Company, Tennessee's Senator Howard Baker and others warned the

---

47. *New York Times*, 22 April 1972.
48. Arthur E. Morgan, *Dams and Other Disasters* (Boston: Porter Sargent, 1971); Arthur Maass, *Muddy Waters* (Cambridge, Mass.: Harvard University Press, 1951); George Laycock, *The Diligent Destroyers* (New York: Ballantine Books, 1970); James T. Murphy, *The Empty Pork Barrel* (Arlington, Mass.: Lexington Books, 1974); John Lear, "Environment Repair: The U.S. Army Engineers' New Assignments," *Saturday Review*, 1 May 1971, pp. 47–53.

TVA that it was in danger of losing its traditional constituency.[49]

### Federal Power Commission

The independent regulatory agency whose commissioners are presidential appointees and responsible for regulation of electric power and natural gas industries. Like so many of the regulatory agencies, the FPC has developed close ties with the power companies it is supposed to be regulating. The FPC "seems content to carry out routine activities that deal with a tiny fraction of the electric power industry's problems . . . [and] has made no effort yet to seek a broader mandate." [50]

### Atomic Energy Commission

Another independent regulatory agency. It is mandated to promote the use of nuclear power and at the same time regulate its use to guarantee safety. The energy crisis of 1974 led to larger budgets for the AEC. Critics have called for the safety division to be split off and made independent. Of its various nuclear programs, its weapons development have received greatest budgetary support.[51]

Though listed here as one among several bureaucratic agencies handling environmentally related policy, the Environmental Protection Agency (EPA) was intended to be the coordinator and monitor of them all. The EPA brought under its umbrella the National Air Pollu-

---

49. Roger M. Williams, "TVA and the Strippers," World, 19 June 1973, pp. 20–41; New York Times, 12 January 1975.

50. David Freeman, "The Forgotten Energy Agencies," Washington Post, 10 February 1974; Michael and Fort, Working on the System, pp. 467–562.

51. Peter Natchez and Irvin C. Bupp, "Policy and Priority in the Budgetary Process," American Political Science Review 67, no. 3 (September 1973): 951–63; Freeman, "Forgotten Energy Agencies"; H. Peter Metzger, The Atomic Establishment (New York: Simon & Schuster, 1972); Michael and Fort, Working on the System, pp. 337–466.

tion Control Administration (NAPCA) from HEW and the Federal Water Quality Administration (FWQA) from Interior.[52] Also incorporated were the pesticide control programs from Agriculture and the Food and Drug Administration and radiation regulation activities from the AEC.[53] Noise abatement was a new office created within the EPA. By 1973 the EPA had 8,000 employees and a budget of billions. Its sewage treatment spending alone stood second only to the highway construction program in total federal public works expenditures. Its research grants were as sought after as those from the AEC or National Institute of Health.

Much of the power to make and enforce environmental policy remains outside the EPA in agencies with longer traditions and more strategic resources, leaving the EPA to be merely one bureaucratic competitor among many. For example, to protect farm workers and consumers from the harmful effects of pesticides the EPA must deal with the production-conscious Agriculture Department, the industry-sensitive National Occupational Safety and Health Administration (NOSHA) in Labor, the often reluctant FDA. The EPA's principal support in protecting farm workers from DDT substitutes was the OEO-funded Migrant Health Project and the courts which forced NOSHA to tighten its safety regulations.[54]

Likewise, to enforce its ban against DDT, the EPA

52. For analyses of the weaknesses of the NAPCA while under HEW, see John C. Esposito, *Vanishing Air: Ralph Nader's Study Group Report on Air Pollution* (New York: Grossman, 1970); Charles O. Jones, "The Limits of Public Support: Air Pollution Agency Development," *Public Administration Review* 32, no. 5 (September/October 1972): 502–8, contains a critique of the FWQA while it was within the Interior Department; David Zwick and Marcy Benstock, *Water Waste Land: Ralph Nader's Study Group on Water Pollution* (New York: Bantam Books, 1971).

53. An analysis of the pesticide-control program is included in Harrison Wellford, *Sowing the Wind* (New York: Grossman, 1972).

54. Daniel Zwerling, "The New Pesticide Threat," *Washington Post*, 5 August 1973.

has had to fend off agencies and interests seeking waivers to the ban. In 1973 it did withstand pressures from the Forest Service, timber companies, and Indian councils in northwestern states which wanted permission to use DDT against the tree-killing tussock moth. But a year later, with pressures mounting as more trees were lost, the EPA relented and granted the waiver.[55]

In both of these fights over pesticide control the EPA has had to be extremely sensitive to the preferences of one Congressman, Jamie L. Whitten, Mississippi Democrat. For Whitten, an outspoken advocate of pesticides, is chairman of the House Appropriations Subcommittee that, until a 1975 shakeup in House Committee chairmen's power, oversaw the EPA's annual budget. If the EPA took too strong a stand on DDT it could have had its budget cut the next year in Congress.

Many interagency conflicts are resolved in the White House. That agency with (1) the most influential clients, (2) the most institutional prestige, or (3) the policy outlook most in harmony with the President's own philosophy will emerge from the White House session the winner. To date the EPA has lost a considerable number of battles when resolution ended in the President's office. For instance, President Nixon shifted the authority to set radiation standards for individual power plants from the EPA back to the AEC, just at the time when the EPA was on the verge of setting strict standards for plants. EPA's supporters feared that this decision would encourage other departments that had lost authority to the EPA to persist in their attempts to regain their bureaucratic territory by appealing to the White House in the name of energy needs.[56] Under the Ford Presidency the EPA found

55. *New York Times*, 29 June 1974.
56. *New York Times*, 12 December 1973, 13 December 1973. Appeal to the White House was less effective during the preoccupation with Watergate. EPA Director Russell Train noted in the midst of the 1974 Watergate proceedings that the White House had "loosened its

itself cast in the role of inflationary villain, forcing businesses to raise prices in order to cover the costs of newly required antipollution devices. To avoid appearing as President Ford's adversary the EPA tried to show that the influence of pollution controls on inflation were infinitesimal.[57]

The chief weapons in the EPA's bureaucratic arsenal have been the NEPA's environmental impact statement requirement and the energetic publicity campaigns and court suits aimed at expanding the EPA's mandate. In addition, the EPA has the advantage of two initial directors with independent statures of their own, and sewage treatment and research funds that permitted the agency to woo sectors of the public which otherwise might not take an interest in the EPA's survival.

Federalism has complicated the EPA's task. Some states have resisted antipollution regulation. As the majority leader of the Arizona House said, "The state would have to pass laws to implement the [EPA] regulations . . . and we are not going to do it." [58] Other states have engaged in delay. Only eight states met the April 1972 deadline for filing remedial traffic plans with the EPA as required under the Clean Air Act of 1970. These delays compelled the EPA to announce drastic traffic-reduction plans of its own for the most congested cities, plans that elicited anger from mayors and city councilmen. The EPA

tight hold" on executive agencies, giving the EPA officials more responsibility and flexibility. *New York Times*, 19 July 1974.

57. This was in fact the conclusion of an independent survey by the Department of Commerce in 1974, when inflation became the uppermost issue. Carolyn S. Konheim, "Letters to the Editor," *New York Times*, 7 October 1974.

58. *New York Times*, 21 July 1973. The important role that states play, through their bilateral bargaining with the EPA, in shaping national environmental policy is underscored in: Robert D. Thomas, "Intergovernmental Coordination in the Implementation of National Air and Water Pollution Policies," paper presented at the Annual Meeting of the American Political Science Association, Chicago, 29 August–2 September 1974.

was placed in the unhappy position of being a heavy-handed federal agency that could not understand grass-roots realities. The EPA had to shelve most of its metropolitan traffic plans.[59]

## CONCLUSION

It is a mistake to portray the admittedly crowded American political scene as a Hobbesian "war of all against all." In American politics there is complexity but rarely chaos. Conflicts and fragmentations adhere to quite regular patterns, patterns which leave many groups with less access and influence in environmental affairs than others. For the most part, those interests that give top priority to economic growth find it easier to be heard in American policy circles and to be taken seriously.

Environmental groups in the United States have been able to mobilize perhaps a wider cross section of the citizenry than in most other countries. But they have yet to convince persons in economically marginal situations that environmental regulations will not jeopardize their jobs. Furthermore, mobilization, while impressive evidence of the resources available for citizen participation in American politics, is by itself no guarantee of effective governmental action. Environmentalists have been thwarted by both the limitations on genuine pluralist politics and the excesses of pluralist politics. On the one hand, established networks of mutual dependency between certain specialized intra- and extragovernmental groups have excluded proponents of antipollution and land-use planning measures from congressional and executive policy making. To loosen up these subgovernments, environmentalists have resorted to court actions and

59. Gladwin Hill, "Anti-Pollution Charade: The EPA's Drastic Proposals," *New York Times*, 23 June 1973.

electoral campaigning. The environmental movement has done perhaps more than any other single cause to bring the subgovernments into the light of day and thus to deprive them of their principal advantage: obscurity.

On the other hand, however, environmentalists have been thwarted by the multiplicity of decision-making centers and the fluidity of issues in the American system. To ensure passage and implementation of even a single air pollution law requires an environmental group to be active in fifty states, three branches, and four congressional committees at once. The fragmentation of American policy formation quickly drains all but the best-endowed groups of their resources. If environmentalists succeed in prying open the lobbyist-committee-bureau subgovernments that now determine environmental policy, they will, of course, be adding to these burdens of pluralism.

Conventional descriptions of American politics concentrate on fragmentation and dynamism but usually without reference to one of the most diverse sectors of the polity: the bureaucracy. Environmental politics in the United States have helped to direct more attention to the bureaucracy and to bureaucrats, who, like party leaders and lobbyists, have value priorities, institutional loyalties, survival instincts, and uncertain budgets.

# POLLUTION POLITICS IN THE SOVIET UNION

## CHAPTER 6

## TOTALITARIANISM OR CONVERGENCE

The American model of pluralist politics has to be revised to take account of unequal political resources, closed-policy "subgovernments," and an increasingly influential bureaucracy. Similarly, Soviet experience with environmental control suggests that we modify the conventional model of Russian politics.

The standard wisdom concerning the Soviet political system is not based on pluralist assumptions but on the concept of totalitarianism. Yet, this model fails to account for some of the most important characteristics of the Soviet policy process.

In its purest form, totalitarianism includes:

**190**

1. permeation of political authority throughout all of society
2. clearly defined hierarchy of command in which decisions are made at the top and carried out obediently below
3. a single political party which ensures standardization of values and behavior
4. restriction of expression except as explicitly condoned by officialdom
5. politics taking the form mainly of administration with little room for competition or popular initiative

While few observers claim that the Soviet political system fits this ideal perfectly, the custom has been to analyze Soviet politics as if this model portrayed its strongest tendencies. That is, if one starts with totalitarian assumptions about how politics works, one will ask the questions and collect the most pertinent to understanding Soviet decision making and compliance.

Recently, scholars have challenged the utility of the totalitarian model as the optimal starting point for analyzing Soviet politics. Examination of Soviet environmental politics adds fuel to that challenge. The Soviet system is anything but monolithic. There are discontinuities, loopholes, conflicting priorities, criss-crossing authorities. Laws that are passed may be grounded in ambiguous ideological rationales; existing regulations can gather dust because of deliberate delays in implementation; compliance when sought at all may be ineffectual and halfhearted. While there are no well-organized opposition organizations outside the Communist party, there are conflicts between party and nonparty officials about the seriousness of pollution and its proper remedies.

When comparing American and Russian politics one is tempted to slip into an unrealistic dichotomy, as if the only way to compare the two systems were to envision them as embodiments of contrary ideals: totalitarianism versus pluralist democracy. Such a dichotomy may be fruitful in philosophical debate, but can be misleading— even worse, dulling—in empirical investigation. Rather than starting from the notion that one is engaged in an ideological struggle, imagine oneself as an explorer in a relatively unchartered land, armed with a sharp eye and a sensitivity gained from exploring other political systems. Employing this analogy, the Soviet system may or may not appear attractive, but it is certain to appear interesting.

One alternative to the dichotomizing approach is the "convergence" theory. This presumes that, regardless of conflicting ideologies, the Soviet and American systems are becoming more and more alike. The motors behind the convergence are industrialization and the knowledge explosion. Both countries are highly industrialized, dependent increasingly on advances in technology, and committed to national strength based on continuing economic growth. This is not simply economic determinism; it presumes that elites in each nation are confident of man's ability to harness nature for human goals.

The emergence of the environmental issue lends credence to the convergence theory. First, the issue's ascendancy itself testifies to the commonalty of two allegedly contrary systems. Russians may be ill at ease with the issue because it challenges socialist ideological convictions about the virtues of state ownership of means of production. Americans may shy away from the issue since it raises troubling doubts about the self-regulating character of free enterprise. Still, in both countries the most threatening aspect of the environmental issue—and thus the most potent force for suppressing it or treating it lightly—is its challenge to the state's goals of social justice and national power via industrialization.

Moreover, the American and Soviet systems are riding the same track leading to growing bureaucratization. In each system are interests and rivalries in the government structure where it was presumed there was mere administration. In addition, as American commentators have come to acknowledge the serious limitations on the free play of interests, so Soviet observers have come to see that there is more room for interest competition in a one-party state than had been presumed. Nowhere is bureaucracy more prominent and intrabureaucratic rivalry more central than in environmental decision making.

The convergence concept has been vigorously debated by American and Soviet intellectuals. Americans who are skeptical point to the persistence of ideology and class differences in the United States despite technological advances. Soviet critics have labeled convergence theorists—*konvergenty* as they are called in Russian—as threats to established patterns of power. The *konvergenty's* opponents contend that such a notion underestimates the continuing class exploitation in the United States and confuses modern management techniques with genuine public control of production.[1]

Indeed, the convergence image should not be overdrawn. Examination of environmental politics does suggest that the American and Soviet systems share an ideological-cultural reluctance to confer priority on environmental issues as well as flawed implementation of environmental laws that exist due to bureaucratic fragmentations. On the other hand, critical differences do remain: American environmental groups have access to more political resources with which to overcome that official reluctance and to monitor hesitant bureaucratic agencies; the Soviet government has greater capacity for

---

1. Donald R. Kelley, "The Soviet Debate on Convergence of American and Soviet Systems," *Polity* 6, no. 2 (Winter 1973): 174–96.

long-range environmental planning when and if environmental problems gain political primacy; conflicts of interest between regulator and regulated are more integral to the Soviet system, even if they do exist in the United States to a greater degree than civics books imply; even acknowledging the American abuses of cost-benefit analysis, Soviet economic theory and accounting practices make it harder to specify environmental costs in a way that clarifies the consequences of certain policy choices for Soviet political elites.[2]

There had been warnings about the need for conservation of Soviet resources since the 1950s. But for two decades the response was piecemeal and unsystematic. Only in the spring of 1971 did the question of environmental control achieve official status. It "arrived" when it was included in the Directives for the Ninth Five-Year Plan for 1971–75. In his report to the 24th Congress of the Communist party, Secretary-General Brezhnev urged that technological and economic advances be combined with a concern for natural resource conservation and air and water pollution.[3] The following year the Soviet leadership affirmed this issue recognition by entering into a summit-level agreement between President Nixon and Secretary Brezhnev for U.S.–Soviet cooperation in matters affecting the environment.

At the same time that pronouncements were being made at the top of the Soviet hierarchy, the Communist party was beginning to mobilize its resources to attack pollution and polluters. Among the laws that the party has sponsored since the late 1960s have been land legislation (1968), public health regulations (1969), and water legislation (1970). Policy was being crystallized and

---

2. The most thorough critique of the Soviet pricing system's inherent bias against environmental protection is in Marshall I. Goldman's *The Spoils of Progress* (Cambridge, Mass.: MIT Press, 1972).

3. Margaret Miller, "Environmental Crisis: The Soviet View," *World Today* 29, no. 8 (August 1973): 352.

legitimized, but enforcement remained in doubt.[4] The weakness of implementation had undermined the earlier Soviet laws dealing with conservation and pollution. For instance, there was the 1960 Russian Republic law, containing general directives on the conservation of land, mineral resources, surface and underground waters, forests, and the atmosphere. The law required new factories to gain permits before discharging pollutants and restricted logging operations. But it wasn't until the 1970s new Principles of Water Legislation that meaningful sanctions were available to ensure the effectiveness of the earlier protections.[5]

Likewise, Soviet air-quality standards had been on the books since 1949 and were revised in 1963. Laws were passed at the republic level between 1957–63 directing factories and town governments to meet these standards. In recognition of the growing number of automobiles, especially in cities such as Moscow, auto-emission standards were established in the 1960s as well. Once again, however, the regulations were poorly enforced and hindered by cross-cutting administrative responsibilities. Only when the political leadership, acting through the party Central Committee or Politburo, selected a particular environmental problem—such as pollution of the enormous Lake Baikal—for special attention did existing laws have their intended force.[6]

4. David E. Powell, "The Social Costs of Modernization: Ecological Problems in the USSR," *World Politics* 23, no. 4 (July 1971): 618.

5. Donald R. Kelley, Kenneth R. Stunkel, Richard R. Wescott, "The Politics of the Environment: The USA, USSR, and Japan" (Paper presented at the International Political Science Association Meeting, Montreal, 1973), p. 15.

6. Siberia's Lake Baikal, the world's largest freshwater body, has been the center of controversy because of government-run cellulose plants operating on its shores and polluting its waters, famed for their purity. As a result of pressure from Soviet scientists, local residents, and international environmentalists, the Russian government belatedly installed waste-treatment facilities in its factories. Some observers contended, however, that the lake continues to deteriorate due to the

Thus observers considered it significant that in the September 1972 meeting of the Supreme Soviet, the Russian legislature, the agenda was devoted entirely to environmental questions. V. A. Kirillin, deputy chairman of the Council of Ministers, gave a keynote address in which he lauded the advantages of socialism in protecting the environment, but admitted that these advantages were not being fully exploited. Kirillin noted that Soviet problems were as serious as elsewhere. They included a doubling of atmospheric pollution in the previous decade, noise pollution that was reaching hazardous levels in industrial areas, seepage of agricultural fertilizer chemicals into waterways. Later speakers in the session went on to criticize bureaucratic red tape and rivalry that aborted antiprograms. Still other speakers questioned whether the current official pricing system accurately reflected the social costs of production.[7]

As significant as this meeting of the Supreme Soviet was in reflecting the elite's growing awareness of environmental problems, it was only three months later, in December 1972, that the Supreme Soviet's agenda had reverted to its usual preoccupation with raising levels of production and accelerating economic growth.[8]

## SOVIET ISSUE CREATION

Issue creation in the Soviet Union is not solely a matter of designation from the top, as the monolithic,

expanding industrial operations. See Marshall Goldman, "Our Far Flung Correspondent: The Pollution of Lake Baikal," *New Yorker*, 19 June 1971, pp. 58–66; Jon Tinker, "What's Happening to Lake Baikal," *New Scientist* 58, no. 850 (14 June 1973): 694–95; "World Environment Newsletter," *World*, 27 March 1973, p. 48; *New York Times*, 18 April 1971, 2 April 1972, 20 May 1973, 27 August 1973.

   7. Kelley et al., "The Politics of the Environment: The USA, USSR, and Japan," pp. 16–17.

   8. Miller, "Environmental Crisis," pp. 356–67.

hierarchical model of totalitarian systems would have it. But official sanction of a policy question is critical to its gaining the status of a legitimate issue. Without this legitimation it is difficult and even risky for persons anxious to draw attention to the issue to solicit support. In this sense, the issue of the environment has the advantage over the issues of, say, ethnic nationality discrimination or the issue of Jewish emigration, both of which are deemed by the party and governmental elites still to be nonissues and thus a challenge to authority if they are promoted as real issues.

Despite the political elite's ability to confer or withhold legitimacy on an issue, independent forces have helped bring the environmental issue to the fore sufficiently that this elite must take some public stand on it. Three of the most important forces have been (1) ad hoc groups of citizens at the local levels who respond to environmental disruptions that affect their own lives; (2) scientific and technical experts who are in roles that make them conscious of the levels of pollution or environmental disruption and equip them to measure the impacts of those hazards on public health and natural resources; (3) mass media controlled by the government (with the exception of the *samizdat*, underground dissident newspapers) but which from time to time publish citizen complaints against officials. What characterizes these three forces and the issue-creation dynamics that together they have set in motion in the area of environmental matters is their essential ad hoc quality. That is, when issues are first emerging and have yet to be coopted by the political leadership, their survival depends on their not appearing to be fundamental threats to Marxist ideology or the power structure.

In every system the environmental issue has grown out of discrete events, piecemeal events, partial awareness. It has taken time to crystallize to the point that it has coherence and nationwide relevance for a wide assortment of otherwise diverse sectors of society. What is distinctive about issue creation in the Soviet system, nevertheless, is that the piecemeal, ad hoc quality is not simply a frustration to be endured; rather this ad hoc quality is a protective covering for the issue, permitting it to grow without appearing to be a threat to elites and thus being squashed before it gains widespread saliency.

The environmental issue also has had to overcome the public commitment to productivity and now to consumer well-being. If officially acknowledged, environmental hazards might put the brakes on production. Yet it was assumed that, given the socialist pattern of state ownership, industry by definition *could not* be injurious to the common man.

This commitment to productivity dictates criteria for measuring officials' or plant managers' success. It is hard for a plant director in the Ukraine to take his polluting effluents seriously if he knows that his future career will be determined not by measurements of the purity of the local river but by his record in meeting production quotas laid down in his central ministry according to the overall Five-Year Plan. Likewise, the public health inspector responsible for imposing environmental standards is all too conscious that his job is deemed less crucial to Soviet greatness than is the job of the factory manager he is supposed to discipline.

Pressure on middle- and lower-level Soviet officials to meet productivity quotas is a persistent theme in Russian newspapers. There are frequent reports of industrial personnel who succumb to bribery and other forms of corruption in order to be certain that their factories will fulfill their quotas. For instance, *Pravda* printed a story of a woman in charge of sales at the Chelyabinsk Tractor Plant. One of the regular purchasers of Chelyabinsk tractors was the Berdyansk Road-Building Machinery

Plant. When the tractor factory ran short of parts and began telling its customers that it would not be able to fill all their orders, the assistant director of the road-building machinery plant became anxious, for his career depended on getting the necessary tractors with which to fulfill his plant's own quotas under the central plan. So the assistant director started paying regular monthly bribes to the woman sales manager; she in turn made sure that the road-building plant received special preference in the distribution of scarce tractors. *Pravda*'s moral was: "Greed should certainly be punished. But those who incited [the sales manager] to crime deserve no less punishment. No matter how much they justified themselves with their concern for the interests of production, a bribe is always a bribe." [9] *Pravda* slapped the culprits' wrists but stopped short of raising more basic questions about the dangers of commitment to productivity distorting social values.

Career patterns shape issues. Persons take seriously those issues that match the criteria by which they are evaluated for career promotions. In the Soviet Union both ministerial officials and state-owned plant personnel advance their careers by giving priority to production over all other concerns. Trade-union officials are similarly affected. In the United States, union officials have been reluctant to acknowledge environmental concerns because they saw their own union jobs dependent on job security and wage increases for their rank and file and those in turn were seen to depend on economic expansion, not increased regulation. It has only been in unions such as the recently reformed United Mine Workers Union and progressive unions such as the United Auto Workers and the chemical workers' union, whose mem-

---

9. "The Expeditor," in *The USSR Today: A Soviet View*, ed. Jan A. Adams et al. (Columbus, Ohio: American Association for the Advancement of Slavic Studies, 1972), p. 27.

bers have suffered from industrial health hazards, that the separation of union role and commitment to economic expansion has been sufficient to permit environmental awareness. In the Soviet Union the trade-union roles and productivity are even more firmly wedded. According to an authorized speech printed in *Pravda*, the "main sphere of the trade unions' activity is the economy and concern for accelerating scientific and technical progress and increasing labor productivity." [10] Thus a union official will be judged by his superiors on the basis of his rank-and-file's productivity, not their health or affluence.

Two recent shifts in Soviet policy priorities add new dimensions to this preoccupation with productivity. Each shift could enhance the status of environmental issues. The first shift is still subject to top-level debate in party and ministerial circles. It involves the movement toward greater emphasis on consumer items and away from producer items of heavy industry. In the Fifth Five-Year Plan of 1971–74 the leadership called for an increase of 7.5 percent in the consumer sector with a 6.6 percent growth in heavy industry.[11] In addition, the leadership's spokesman addressing the people's representatives in the Supreme Soviet pledged a 5 percent growth in real incomes for ordinary Soviet workers as well as speed-up in apartment construction.[12]

One specific repercussion of the shift to consumer priorities has been the increasing production of private automobiles—and thus the increase in air pollution caused by auto emissions. Car owners by 1973 were estimated at close to 2 million in a nation of 250 million, minuscule when compared to the proportion of car owners in other industrialized nations. Still, the future

10. "The Party and the Mass Organization," in Adams et al., *The USSR Today: A Soviet View*, p. 9.

11. *New York Times*, 13 December 1973.

12. Ibid.

promises a marked rise in this figure and thus in the environmental problems. The Italian-built auto plant at Togliatti (named after the Italian Communist party leader) on the Volga River is approaching its designed capacity for more than 600,000 units per year.[13]

Furthermore, although nationally the number of private autos is small, they are concentrated in a few major cities, especially in the western region, and already there are enough cars in Moscow, combined with trucks, to cause regular traffic jams and to prompt intellectuals to wonder whether the Soviet auto owner will be tempted toward a new sense of individualism.[14] *Izvestia* printed a report from Kazakhstan that auto-emission inspectorate stations were being set up. The persons taking the lead in devising new auto-exhaust devices and air-quality standards were scientists in the Kazakh Republic Academy of Science. The report concluded optimistically that soon citizens in the Kazakh capital of Alma-Ata would witness street scenes like this: "A state motor vehicle inspector raises his black-and-white traffic baton and demands, 'Let me see your exhaust toxicity certificate.' Should the certificate lack the stamp of an inspection station, the position of the driver would be anything but enviable." [15]

For this prophesy to be realized, however, there have to be persons attracted to such an inspectorate and inspectors confident enough of their authority and the support of their superiors to enforce such standards. Ultimately these conditions will be determined not by scientists in the Kazakh Academy, but by the priorities of the Ministry of Motor Vehicles. In the area of noise pollution, for example, the minister of public health launched a complaint that his counterparts in the Minis-

13. *New York Times*, 25 September 1973.
14. Murray Seeger, "Wheels for Ivan," *Washington Post*, 18 June 1972.
15. "Fighting Auto Pollution in Kazakhstan," in Adams et al., *USSR Today*, p. 124.

try of Motor Vehicles had not even devised official standards for motor noise levels, even though public health officials had found noise levels exceeded health standards on major motorways in Moscow, Volgograd, and a number of other cities.[16]

Thus while Soviet policy-makers' growing sensitivity to the living conditions of average citizens could engender a greater awareness of the social costs of productivity, the connection is not automatic. Consumerism and environmentalism are not mutually exclusive: neither are they inevitable allies.

The second shift in top-level policy has been the promotion of modern management techniques. Once again, the change does not lessen the commitment to productivity, but it might alter the terms in which production policies are evaluated. Observers who foresee the convergence on the Soviet and American systems pay close attention to the managerial reforms following from more sophisticated technology. The most visible artifact of these reforms is, of course, the computer. Soviet reformers are critical of the Russian lag in both the development of computer technology and its application to economic management. To catch up, the Soviet leadership has promoted technological exchanges with the United States. Not surprisingly, these moves elicit wariness from middle- and lower-level managers whose training has ill equipped them to handle this new mode of operation.

The deputy minister of foreign trade, for example, chastised Soviet industrial personnel for trying to avoid trade with the Common Market, Japan, and the United States because it was too demanding and competitive. Borrowing capitalist technology would not hurt commu-

---

16. "Noise Pollution in the City," in ibid., pp. 124–25.

nist ideology or "humiliate our dignity," he assured. The minister then made unflattering comparisons between Soviet industrialists and those of such countries as Japan, the United States, West Germany, and even Brazil. He pointed to Soviet auto producers' tardiness in adapting technology to changing conditions. By contrast, West German car manufacturers were attentive to new American regulations and were quick to adapt their cars to the new antipollution controls. At bottom, the minister argued, the Soviet export program was "hobbled by inefficiency, bureaucratic delays, an attitude among industrialists that exporting is punishment and not an opportunity, and endless bickering among agencies and enterprises aimed at avoiding responsibility for exports rather than trying to stimulate them." [17]

So long as the environmental issue appears to be in conflict with the drive for higher rates of productivity, it is at a distinct disadvantage in Soviet politics. It may gain legitimacy by slipping in a side door, thus avoiding direct confrontation with this traditional policy priority. The growing concern expressed by the Brezhnev regime in consumer demands and in managerial and technological innovation do not automatically promote environmental interests, but they might allow it access to major policy forums. First, the consumerist argument moves party and government leaders to take more account of the impact of five-year plans on the everyday lives of ordinary citizens. This, in turn, permits discussion of the relationship between GNP goals and the quality of life, a principal element of which is a healthy environment. Second, the drive for managerial and technical innovation encourages

---

17. From an article by Nikolai N. Smelyako in the December 1973 issue of *Novy Mir*, reported in the *New York Times*, 5 February 1974. See also a three-part series by Henry Lieberman on Soviet Management Innovations: *New York Times*, 12, 13, 14 December 1973.

exchange with economies outside Eastern Europe, many of which have been ahead of the Soviet Union in environmental awareness and technological adaptation to meet new environmental standards. These strictly economic policies may give the needed underpinning to the cooperative machinery set up between the United States and the Soviet Union in their 1972 treaty.[18]

While there has indeed been a marked rise in the issue's visability, it has not been due to mobilization of nongovernmental forces pressing new demands on officials. Local citizens entering specific complaints have had an effect when their letters are published by government organs. Scientists have stimulated interest in the issue because of their special skills and their privileged and less vulnerable positions in Soviet society.[19] Underground dissidents and their more daring public spokesmen speaking through unofficial papers or bold letters to party leaders have linked criticisms of pollution to larger issues of Soviet development and liberalization. But the decisive factor remains the central leadership in the upper ranks of the party and government and their policy priorities. It is their pronouncements that make or break an issue.

If the environmental hazards become a matter of concern, for instance, in the siting of one of the world's largest steam electric stations in the Ukraine, it is because economic planners are taking a greater interest in those

---

18. Among the projects cited for joint American-Russian research were analysis of the urban environmental problems of Saint Louis, Missouri, San Francisco, Atlanta, the Delaware and Potomac rivers, and the San Andreas Fault in California in the United States; the earthquake regions of the Pamir Mountains, Lake Baikal, and Leningrad in the USSR. See *New York Times*, 22, 23 September 1972.

19. For example, see Philip R. Pryde, "Soviet Pesticides," *Environment* 13, no. 9 (November 1971): 16–24. Likewise, Soviet zoologists took the leading role in calling for government protection of the nation's wolves. *New York Times*, 11 February 1973.

matters, not because the plant has stirred up local residents in opposition.[20] If the Soviet public hears belatedly about a break in one of Russia's biggest oil pipelines near the Caspian Sea, it is because the editors of *Pravda* decide to publish the story and couple it with a criticism of the Ministry of the Petroleum Industry for its delay in reacting to the emergency.[21] When major clean-up campaigns are launched it is because of special promotion by party and government elites, as in the case of the recently announced antipollution campaign to revive the Volga River.[22]

In assessing whether environmental disruption has become a salient political question in the Soviet system, therefore, one has to look not at opinion surveys (though they are being used more widely now in Russia) or at electoral behavior or associational memberships. Rather, one can best gauge the status of a potential issue by looking at (1) the separate mention given to environmental matters in top level discussions of national five-year plans; (2) the level of capital expenditures for soil, water, and air conservation and antipollution in the five-year plans; (3) the criteria used to evaluate plant directors' performances; (4) the willingness of officials to join in international projects dealing with environmental protection.[23]

---

20. *New York Times*, 27 November 1972.

21. *New York Times*, 22 March 1971. In 1973, there was another accident. An offshore oil rig in the Caspian Sea caught fire and burned for several weeks, spewing oil across the water. But the area had become dependent on the oil industry, with 25,000 oil workers, each of whom were earning salaries well above the Soviet average. *New York Times*, 23 June 1974.

22. *New York Times*, 22 March 1971.

23. In the past few years capital investments in soil, water, and air conservation have been growing at a faster rate than capital investments in the national economy as a whole. In the Ninth Five-Year Plan for 1971–75 appropriations for environmental protection exceed investments in such important branches of the economy as railroad transportation. "World Environment Newsletter," *World*, 19 June 1973, p. 35.

## GROUPS AND POLICY PROCESS

Communist political systems are characterized by parallel structures of authority: the organs of government and the organs of the Communist party. Relations between these parallel structures are not identical in each communist system, and some of the most exciting research in political science is now being done in comparative communist systems.[24] But in the Soviet Union, as in most other communist systems, the highest organs of the party—the Politburo and Central Committee—wield the greatest influence in policy decisions. The simplistic notion of a dictational process of commands is less accurate than the picture of informal debates and bargaining among top party officials. The government structure is headed by many of these same party leaders wearing different hats. The Council of Ministers is roughly equivalent to the British or Japanese cabinet and is composed of the ministerial chiefs of the multitude of government ministries.

Environmental interests lack power within this structure. In the Council of Ministers it is the heads of the ministries concerned with *heavy industry* who carry particular weight. Environmentally oriented state agencies and semiofficial conservation groups "are usually closed out of these high-level negotiations because of long-stranding dominance of party *apparatchiki* (career personnel) and industrial interests." [25]

Interplay of interests exist in the Soviet policy process but not in forms familiar to Americans. Interests are pursued; there are rivalries over personal influence and

24. See the journal *Studies in Comparative Communism* published by the University of Southern California.
25. Kelley et al., "The Politics of the Environment: The USA, USSR, and Japan," p. 9.

policy claims; power is distributed unequally among various sectors of the policy-making elite and subelite. But rarely are interests embodied in formal organizations outside those structures created to serve the state. Rather, interests are furthered through informal networks of communication among Russians in the same ministry, or in the same occupation, or in the same political generation.[26] At times an analyst must be cautious and not see an interest group under every bed. What may look like a group sponsorship of some reform or project may in fact be a coincidental convergence of policy stands by individuals who share certain common characteristics.

This is true, for example, in the influence wielded for the sake of heavy industry and the pursuit of productivity above all other policy goals. On the one hand, officials in the ministries of petrochemicals or motor vehicles have organizational career stakes in promoting policies that fatten their budgets and make their programs more feasible. These officials, nevertheless, share common skills and educational training, largely in engineering. Without any informal meetings or phone conversations they might end up on the same side at the bargaining table because of the value they have been taught to place on physical output. Added to this natural inclination among men of similar backgrounds in the government ministries are the backgrounds of party officials. Studies have found an increasing tendency for top-level party posts to be filled by men who have had technical training, many in engineering.[27] Many studies point to the prob-

---

26. H. Gordon Skilling, "Groups in Soviet Politics: Some Hypotheses," in *Interest Groups in Soviet Politics*, ed. H. Gordon Skilling and Franklyn Griffiths (Princeton, N.J.: Princeton University Press, 1973), pp. 29–33.

27. John A. Armstrong has found that the emphasis on engineering among state administrators is true for both the czarist and communist era in Russian history. John A. Armstrong, *The European Administra-*

lems that the Communist party has in a technological age of recruiting persons equipped with sophisticated skills while still maintaining the party's ideological dominance. The principal recruiting strategy appears now to be one of recruiting persons who already have had careers as technicians into party posts, hoping to socialize them into being loyal party members, not simply ambitious technical experts. The division between the career *apparatchiki*, who have built their careers solely within the party bureaucracy, and the coopted technical specialists is a principal line of cleavage of many policy issues.[28]

In addition to the problem of reconciling party recruitment with demands for technical specialization, there has been the ongoing debate within the Soviet leadership over the relative merits of centralized planning and decentralization. Centralization of decision making reached its zenith during the Stalinist era. Recently, there have been deliberate efforts to decentralize decision making in order to foster flexibility and to maximize the talents of middle- and local-level officials. But environmental control has fared poorly when the local officials, especially plant managers, expand their authority, for they have been loath to spend their budgets and labor resources on antipollution practices that do not promote productivity. On the other hand, decentralization could give more influence to those environmentally conscious officials who are more common at the middle ranges of the regime than at the top.

Although informal groups and networks play an important role in Soviet decision making, there are formal semigovernmental organizations explicitly devoted to environmental affairs, especially when defined in terms of

*tive Elite* (Princeton, N.J.: Princeton University Press, 1973), pp. 178–79, 317.

28. Michael P. Gehlen and Michael McBride, "The Soviet Central Committee: An Elite Analysis," in *Man, State, and Society in the Soviet Union,* ed. Joseph L. Nogee (New York: Praeger, 1972), pp. 113–29.

conservation. There is a "plethora of uncoordinated, largely advisory bodies, supplemented by the 20 million persons organized in seemingly important 'public nature protection committees' under the auspices of the Central Council of the All-Russian Nature Protection Society." [29] This group is limited not only by its lack of coordination, but also by its reliance on government sponsorship. Furthermore, the society's large youth section is made up largely of schoolchildren.[30]

Soviet environmentalists have urged that a single all-union agency be created to enforce environmental standards on other ministries. Such an agency could provide a unifying focus for the now fragmented environmentalists themselves. Although the environmental groups have had little influence on the national level, they have had more impact at the local level, especially in conducting semiofficial inspection programs and reporting pollution violations to appropriate state agencies. Local volunteers have used the media in their campaigns and may have thus had indirect influence on the Central Committee members through these vehicles.[31]

Scientists, too, have shown skill in using the media to publicize their environmental findings and perhaps bringing them to the attention of national policy makers. In addition, national laws now encourage environmentally responsible services within the Ministry of Public Health to create regional advisory commissions on which local industrial, governmental, and environmentalist representatives are appointed. As these agencies play a larger role in advising the central ministries, they provide environmentalists, many of them scientists, an avenue for policy influence.

29. Keith Bush, "Environmental Problems in the USSR," *Problems of Communism,* July/August 1972, p. 28.
30. Kelley et al., "The Politics of the Environment: The USA, USSR, and Japan," p. 10.
31. Ibid., pp. 10–11.

The environmental issue also gives local soviets (local representative bodies) an opportunity to increase their own importance insofar as the soviets can take the part of ombudsman for their regions.[32] One study of the backgrounds and institutional affiiliations of individuals who have contributed environment-related articles to the general media found that scientists and academicians composed the largest single group. But, in addition, in sizable proportions were officials of conservation and pollution-control agencies, state officials in nonproduction agencies such as public health, and the deputies to the national and local soviets.[33]

Two of the best-known studies of environmental policies in the USSR, one by Marshall Goldman and the other by Philip Pryde, decry the absence of environmental pressure groups in the Soviet Union and the resultant lack of autonomous public involvement in policy making.[34] When compared with the United States, Japan, and Britain, the lack of such independent and explicitly organized groups is strikingly evident. There are two possible responses to this. First, while there are no independently organized and legitimized environmental groups equivalent to Italy's Nostra Italia or the U.S.'s Sierra Club, clusters of interested people still exist, some of them with access to the media, ministries, or local governments, who do play the parts of environmental advocates. And within the formal hierarchies of government and party are cleavages over policy and power which do permit the intrusion of influences, especially by technical specialists, on certain policy questions.

32. Donald R. Kelley, "USSR: Room for Specialists," *Environment* 15, no. 8 (October 1973): 26.
33. This study is part of John Kramer's "The Politics of Conservation and Pollution in the USSR" (Ph.D. dissertation, University of Virginia, 1973).
34. Goldman, *Spoils of Progress*; Philip Pryde, *Conservation in the Soviet Union* (Cambridge, England: Cambridge University Press, 1972).

A second response comes from orthodox Marxists. For instance, William Mandel, an American Soviet observer and defender, objects to the Goldman and Pryde analyses. Mandel contends that in fact the Soviet Union has moved effectively in the last decade to reduce environmental disturbances. But then he goes on to make more specific comparisons between the environmentalist movements in the United States and the Soviet Union. He cites three major features distinguishing the two:

1. First, there is the absence of private ownership interests whether in oil or other mineral rights, marketable stands of timber, seacoast, lake frontage, farmland, etc.—or in the profit-motivated manufacturer of air-polluting types of automobiles.

2. There is the psychological effect resulting from everyone thinking of everything as his business . . . a mass popular upsurge developed to protect Lake Baikal because literally *the people* of the Soviet Union believe that nature must be conserved.

3. The third difference between the two ecology movements is in the level of mass participation. Moscow is a city of the same population as Los Angeles, and smaller than New York. But in April of last year 2,000,000 Muscovites turned out for a day of clean-up and beautification, very largely consisting of planting trees, as well as bushes and flowers. No such degree of participation is conceivable in any American city.[35]

One should avoid, as Mandel suggests, drawing comparisons of interest-group politics from country to coun-

35. William M. Mandel, "The Soviet Ecology Movement," *Science and Society* 36, no. 4 (Winter 1972): 395.

try without supplying the broader context in which politics occurs. On the other hand, Mandel's analysis ignores the interests that public officials have, interests which may not be identical with those of an Exxon executive in pushing for offshore drilling, but nonetheless interests that can assign more weight to his own organization's well-being and his personal career aspirations than to some vaguer notion of the Soviet general good.

## BUREAUCRACY

Environmental politics in the USSR is essentially bureaucratic politics. There is a tendency to think of bureaucracies as including only government administrative services. But in the Soviet system not only the careerists within the various ministries are subject to the patterns of bureaucratic behavior; so too are the personnel who staff the parallel structures of the Communist party, the *apparatchiki*. Within the bureaucracies is the added problem of distinguishing between those men who perform conventional jobs in a clearly defined hierarchy and those persons who have considerable freedom to exercise their own judgment. The standard model of a bureaucracy in an hierarchical state implies that bureaucrats are cautious and conservative, that they are essentially passive except when they are pressed to be innovative, at which point they will actively resist change. According to this model, the Soviet bureaucrat is little more than a faceless cog in a giant machine. Currently, however, studies are revealing that Soviet bureaucrats in some areas are capable of taking the initiative and that the policies of managerial reform and decentralization may promote this activist pattern.[36]

36. Jerry F. Hough, "The Bureaucratic Model and the Nature of the Soviet System," *Journal of Comparative Administration* 5, no. 2 (August 1973): 134–68.

Given this trend, fragmentation within the Soviet governmental bureaucracy will be exacerbated. For initiative and innovation taken on behalf of one's own ministry or industrial plant create formidable obstacles to any national environmental regulations.

Referring to the narrow "departmentalism" that characterizes the Soviet system, one Russian observed:

> Each department can defend its own interests and put forward its own reasoning. But there is not one agency among them which is wholly responsible for the conservation and augmentation of our national resources.[37]

Such organizational parochialism was displayed during the 24th Communist Party Congress when the minister of agriculture accused the minister of power and electrification—both Ukranians—of needlessly flooding thousands of hectares of good farming land for his own ministry's power stations. The power minister had not thought to communicate, much less coordinate, his plans with other ministries.[38]

While on paper political power and policy are centralized in the hands of the party Politburo, in practice legal and administrative responsibility is divided between parallel national and union republics—ministries—much as it is in the American federalist system—with questions of pollution abatement and resource conservation left to the latter. Union republic legislation closely follows guidelines set down in Moscow—more so than in the American federalist system—but the existence of parallel and overlapping agencies creates confusion and bureaucratic ineffectiveness.[39]

37. Quoted in Bush, "Environmental Problems in the USSR," p. 28.
38. Ibid. See also *New York Times*, 9 April 1971.
39. Donald R. Kelley, "Environmental Policy Making in the USSR" (Paper presented at the Northeastern Political Science Association Meeting, Saratoga Springs, N.Y., November 7–9, 1974).

Then there is the bureaucratic problem of role conflict. One recalls that it was a Russian who noted that in much of the environmental policy area the "goat is left to guard the cabbage." In the Soviet administrative structure "almost every agency charged with preservation of some aspect of the environment is itself involved in the exploitation of natural resources." [40] The consequence is a large number of bureaucratic situations akin to that of the U.S. Army Corps of Engineers or the internally schizophrenic U.S. Department of the Interior. For example, water resources are supposed to be cared for by the Soviet Ministry of Land Reclamation and Water Resources along with its ministerial counterparts at the republic level. However, the ministries of Power and Electrification, Fisheries, Agriculture, Inland Water Transport, and Public Health also play a vital role and water resource protection is often more a matter of how much bureaucratic clout a ministry has rather than how vital a water project is to national interests.[41] Likewise, the same Land Reclamation officers who are charged with protecting the Soviet Union's precious freshwater resources are also charged with planning for such ambitious public works projects as altering the flow of major northern waterways so that the southern regions have greater access to water, an undertaking that many environmentalists argue will irreparably damage the valuable Russian fish resources.[42]

Besides conflicts and organizational rivalries within the Soviet bureaucracy, weak implementation of laws and

40. Bush, "Environmental Problems in the USSR," p. 28.
41. Kelley et al., "The Politics of the Environment: The USA, USSR, and Japan," p. 13.
42. "Plan to Direct Northern Waters to South," in Adams et al., "USSR Today," pp. 129–30. Similarly the Soviet Ministry of Forestry is responsible for conservation, but is also a prime exploiter of Soviet timber in sparsely forested and relatively developed areas. "Environmental Newsletter," *World*, 10 April 1973, p. 36.

ineffectual imposition of penalties on polluters, most of whom in a socialist system are themselves state officials, have retarded environmental protection. The difficulties of state agencies disciplining other state agencies is not peculiar to a socialist state. But the problems are acute in a system where the means of production are controlled by the state.

At the lowest level, inspectors face a complex monitoring task yet possess little social or political prestige and are equipped with vaguely written laws, ineffectual penalties, and potential culprits who usually have influential backers in their own ministries. Inspectors, moreover, rarely can count on organized citizen pressure on the plant manager or ministerial engineer to supplement his own official pressure, and cannot rely on activist courts to support administrative sanctions. Often the absence of citizen groups and court interventions is a relief to inspectors, who see such intervention merely as a threat to their bureaucratic insulation.

Air pollution standards compliance is put in the hands of sanitary inspectors within the Sanitary Epidemiological Service, an agency of the Ministry of Public Health. Legally, the inspectors are empowered to impose fines on offending factories or municipal agencies. In extreme cases they can even order a halt in production or a relocation or closing of polluting enterprises. In practice, those heady powers fade. "When pitted against powerful production ministries working in close coordination with local officials intent upon regional economic development, the sanitary inspector usually finds himself out-maneuvered and out-lobbied." [43] One example of the

---

43. Kelley et al., "The Politics of the Environment: The USA, USSR, and Japan," p. 14. They are also "out-paid" and "out-tamed." Within an industrial plant, personnel assigned to water-disposal shops have salary levels lower than in other sections of the plant. They also have fewer opportunities to earn production bonuses and receive lesser retirement benefits. It is perhaps not surprising, then, that "large

facile maneuvering with which inspectors must cope should sound very familiar to American environmentalists. A report in *Izvestia* began as follows:

> They say that every summer the Black Sea fisherman swallows several kilograms of sand along with his fish soup, fried fish, vegetables and fruit. He feels quite well, though, and is not inclined to complain to anyone.
> As for the residents of Kuvasai, Fergana Province, they are unhappy; naturally, they complain. First of all, they are not fishermen. Secondly, it is not coastal sand they are swallowing but cement dust. Every day the local combine showers dozens of tons on the small town. . . .
> The residents of Kuvasai, one could say, are fed up with the situation. They insist that the combine install filters to trap the dust and that it stop polluting the atmosphere.[44]

The effort to improve Kuvasai's air became a struggle between sanitary inspectors and the combine's directors. The inspectors seemed to win the first round: the directors installed electric filters on the combine's smokestacks. But they lost the critical second round. For the filters required maintenance and extra labor, both added expenses for the plant directors. So they decided to mollify the local citizenry by operating the filters during the daytime. But during the night the filters were turned off and Kuvasai's sky was allowed to turn gray. As the *Izvestia* reporter remarked, "Crafty, to say the least!" [45]

The inspectors were not naive. They discovered the director's devious avoidance of the law and demanded

---

numbers of anti-pollution specialists at the Baikalsk Pulp Plant near Lake Baikal give up their jobs every year to move to Bratsk, where they receive higher wages at the timber industry complex." Powell, "Social Costs of Modernization," p. 624.

44. "Pollution," in Adams, "USSR Today," p. 127.
45. Ibid.

that the filters be turned on twenty-four hours a day. With that the directors "ordered the guards to throw the sanitary inspectors out and keep them off the premises."[46] The inspectors then went to the third round, where success remained elusive. They appealed to higher sanitary authorities. But their appeals stopped with the provincial officer who replied, "But I cannot summon the director of the combine . . . Comrade Ganiyev is an influential man in his district."[47]

Even when fines are imposed, they have proved ineffective in reducing polluting practices. First, Soviet municipal governments use fines collected from offenders to finance construction of local parks and clubs, thus creating a certain "civic state" in continued pollution.[48] This situation is not unlike that in the United States in which gun-license fees are earmarked for park development, thereby cementing a strange alliance of conservationists and antigun-control lobbyists. The utilization of fees or fines can make strange bedfellows in Soviet as well as American politics.

Second, the accounting methods used in Soviet industries deflate the impact of antipollution fines. The fines are small, usually 100 rubles (about $122). They are paid out of the enterprise's own funds and thus not a penalty felt by the manager personally.[49] Fines have done little to reverse production and environmental priorities; production quotas remain number one.[50]

---

46. Ibid.
47. Ibid.
48. Bush, "Environmental Problems in the USSR," p. 29.
49. Ibid.
50. For an analysis of the autonomy of Soviet plant managers and the production incentive system that motivates them, see David Granick, "Management Incentives in the USSR and in Western Firms," *Journal of Comparative Administration* 5, no. 2 (August 1973): 169–99. Powell also notes that during 1960–64 "more than 25 percent of the funds allocated to plants for the construction of purification installations in all industries was not put to use." Powell, "Social Costs of Modernization," p. 625.

Courts have been an important instrument for altering bureaucratic behavior in the United States and other nonsocialist industrialized countries. In the Soviet Union, since 1969 there have been a few instances in which courts have meted out stiff jail sentences on pollution law violators.[51] But in more routine administrative processes, Soviet leaders have been hesitant to use the courts to pressure bureaucrats. Instead, citizen complaints are channeled through newspapers, as in the case of *Izvestia*'s report on the polluters of Kuvasai. The hope has been that public criticism and ridicule will have positive effects on the bureaucrats. Citizens may also file complaints directly with the bureaucratic agency involved, but that agency is not legally bound to act on the complaint, merely accept it.[52]

"Bureaucratism" has been a concern of Soviet leaders ever since the 1920s. In recognition of its persistent obstruction to effective environmental control, the government in January 1973 announced that it would establish a national environmental protection service to monitor air and water pollution throughout the country. Initially, it appeared that this new centralized agency

---

51. Bush, "Environmental Problems in the USSR," p. 29.
52. Robert J. Osborn, "Citizen Versus Administration in the USSR," Studies in Comparative Federalism, Paper No. 7 (Philadelphia Center for the Study of Federalism, Temple University, n.d.). Furthermore, filing such complaints can be dangerous. A Soviet woman working in Artificial Filament Plant 523 in Ryazan claimed that she was confined to a mental hospital for a month as punishment for informing the Soviet Communist party Central Committee (with a copy of the complaint sent to the United Nations) that her factory had no smoke- or emission-control devices and consequently emitted hydrogen sulphide and other noxious substances into the air, causing a high incidence of pollution-related illness in the area. She was released from the mental hospital only after promising not to write any more letters to international organizations. *New York Times*, 6 August 1974. In a letter defending Svetlana Shramko, exiled Soviet novelist Aleksandr Solzhenitsyn contended that the Artificial Filament Plant was built "simply because the former party secretary of Ryazan Province, Aleksei Larionov, wanted to enhance his own standing in the party hierarchy." Letters to the Editor, *New York Times*, 30 September 1974.

would coordinate fragmented and competing natural resource programs scattered throughout the Soviet administrative structure.

However, the joint government-party decree called on all ministries to take responsibility for pollution abatement in their own policy areas.[53] Thus, from the start the program was fraught with ambiguity: a new centralized agency with nationwide responsibilities but continuance of fragmented pollution programs. Six months after the agency's creation, observers concluded that Soviet policy makers still had been unable—or perhaps unwilling—to permit a strong, centralized environmental protection agency. The basic reason for the continuing ambiguity lies in, first, the persistence of production priorities among the political elites and most of their subordinates and, second, a profound reluctance to admit the limits to which state ownership of the means of production can by itself guarantee that production will be in the people's best interest.[54]

CONCLUSION

It has been a deep-seated devotion to productivity that has given rise to environmental disruption in the Soviet Union, rather than laissez faire economics of mass consumerism. This same preoccupation with productivity has made government and party leaders wary of giving environmental problems the issue status they deserve. On the other hand, reluctance to acknowledge the issue is not fueled by fear of job losses as it is in the United States, since the Soviet Union's problem is labor shortage rather than unemployment.

53. *New York Times*, 10 January 1973.
54. Marshall Goldman, "Pollution Soviet Style," *Business and Society Review/Innovation*, no. 6 (Summer 1973): 44–50.

The production ethic has elevated to top political posts men and women whose training and career aspirations bias them against policies that would restrict industrial output for the sake of public-health protection of nature. Likewise, those central ministries responsible for heavy industry have superior "clout" in policy discussions and can frustrate the operations of agencies assigned to implement environmental laws.

The Soviet Union is not a monolithic totalitarian state, but rather is riddled with competing interests. Yet those interests do not depend on mobilized public support; they are generally confined within the party and state structures. Autonomous citizen groups play only a minimal role in creating environmental issue saliency or pressing bureaucrats to fulfill their environmental duties. Therefore, the regulator is often the regulated. The goat is firmly staked inside the cabbage patch. How effective environmental regulation will be and how seriously it will be treated depends then on how far Soviet political elites go in separating Marxist notions of communal welfare from notions of optimum productivity. So long as the two are presumed to be synonymous, environmental control is likely to be little more than tokenism. That is, if the cabbage patch is not to be devoured, the goat will have to change his tastes.

# POLLUTION
# POLITICS
# IN JAPAN

CHAPTER 7

## "JAPAN, INC."

Japan is luckier than the Soviet Union or the United States insofar as it has not been wrapped in an ideologically based analytical package. Our model of Japanese politics is based more on empirical research with fewer ideological presumptions. Yet the American view of the Japanese political process is influenced by our experiences during World War II and especially during the postwar Occupation. Between 1945 and 1949 Americans, including numerous social scientists, under the command of General MacArthur, set about to remodel Japanese society and government. We started from certain presumptions concerning what made Japan go to war in the 1930s and '40s: family-controlled big business, centralized politics, emperor worship, school-inculcated chauvinism, military political influence. On the basis of these presumptions—some mistaken, others essentially correct but grossly oversimplified—the Americans broke up some institutions, emasculated others, reorganized or created from scratch still others.[1] The antidotes to this intimate involvement in Japanese development have helped to save us from such ideological preoccupations as obscure our study of Soviet politics. Continuing access to Japanese politics has allowed outside analysts to update and revise their earlier hypotheses. Moreover, a younger generation of Japanese social scientists is publishing research unhampered by American preoccupations.

The model of Japanese politics that American and Japanese research has generated goes under two different labels: "factional politics" or "Japan, Inc." While the former is more concerned with the patterns of behavior that mark the workings of political groups, the latter is more concerned with the special relationship between private and public power. Both include essentially similar ingredients:

1. a traditional and persistent orientation of Japanese toward group identification combined with a portrayal of society as a series of hierarchical relationships
2. a Japanese cultural abhorrence of direct interpersonal conflict and belief in the ethics of reciprocity and obligation
3. political attachments based on personal allegiances rather than issues; within the various political parties this produces factionalism
4. political parties that have weak or nonexistent

1. One of the most systematic analyses of the Occupation and its long-range impact is Herbert Passin's *The Legacy of the Occupation* (New York: East Asian Institute, Columbia University, 1968).

grass-roots organizations, relying instead on support from large national organized interests (business or labor groups in particular)
5. a high degree of centralization in authority and decision making, leaving local governments and groups relatively impotent
6. a deeply entrenched alliance between big business and the ruling Liberal Democratic Party, which has its roots in the modernizing strategy of the samurai elites who launched the Meiji Restoration in 1868
7. social prestige and influence for civil servants[2]

"Japan, Inc." refers to a meshing of national values and interests that is now challenged by the antipollution issue. The Meiji Restoration of 1868, which pushed Japan into the modern era, was founded on the belief that a nation able to prevent foreign colonization would have to be able to compete industrially. That, in turn, according to the Meiji oligarchs and their successors, necessitated a strong central government, close cooperation between policy makers, bankers, and manufacturers, and a loyal public whose labor could be mobilized for the sake of fulfilling government-prescribed national goals. The humiliating defeat in World War II subtracted militarism and emperor domination (often more symbolic than real) from this formula. In its postwar form, the formula makes Japanese national welfare and individual happiness synonymous with material prosperity and rapidly rising GNP. It encourages mutually supportive relations be-

2. Some of the most perceptive analyses of Japanese society and culture are George DeVos, *Socialization for Achievement: Essays on the Cultural Psychology of the Japanese* (Berkeley: University of California Press, 1973); Ezra F. Vogel, *Japan's New Middle Class* (2nd ed.; Berkeley: University of California Press, 1973); Bradley M. Richardson, *The Political Culture of Japan* (Berkeley: University of California Press, 1974).

tween civil servants in key economic ministries and
directors of large private firms such as Mitsubishi, Sony,
or Nippon Steel, for the sake of ensuring that material
well-being. There are disagreements, but they are presumed to be solvable. It puts a high priority on government economic planning in which civil servants play as
vital a role as elected politicians. A booklet prepared by
the U.S. Commerce Department for American businessmen doing business with Japanese sets forth the following attributes of "Japan, Inc.":

—When all the exceptions and nuances of the government-business relationship have been taken into
account, "Japan, Inc.," is an economic fact of life.
—There is a special style and scope to interaction
between government and business in Japan which
makes it distinctive; especially its consensual approach and the generally shared desire to advance
Japan's national interests, both of which serve to
overcome conflicts between government and business officials.
—But "Japan, Inc.", is not the Colossus nor the
Conspiracy between government and business that
its most severe critics make it out to be: in contrast
with a monolithic, Japan's system includes autonomous actors who on occasion do follow independent courses of action.
—The government depends more on inducements
than controls to persuade business to follow a more
desired course. The government's direct control
powers have been steadily waning. Among the
inducements the government has in its arsenal are
financing, tax concessions, subsidies, technical assistance. The carrots are larger than the sticks.
—Interaction between government and business is
pervasive in the Japanese economy but not all-
encompassing. The managers of Japan, Incorporated, focus their attention mainly on the growth
sectors of the economy. . . . The degree and success of government-business interaction not only

varies from industry to industry but in time within the same industry.[3]

The formula has had its outspoken critics within Japan. The opposition parties, with one exception (Komeito), are socialist or communist. The universities and intellectual community include prominent Marxists and opponents of materialistic culture. But during the thirty years since the war this formula has restored Japan to international power. Gross national product has soared; per capita income ranks only twelfth in the world but affluence has made color televisions, electric rice cookers, and automobiles common possessions of ordinary Japanese. Unlike the United States, income is relatively evenly distributed; Japan is a nation of a vast middle class with few very rich or very poor.[4] Overseas, Japanese products and investment ventures have shaped the economies of Hawaii, British Columbia, Brazil, South Korea, Taiwan, and scores of other territories. Despite the famed dedication of Japanese to work hard, leisure has become big business, to the point that the *Japan Times* reports that nearly one percent of all land in the nation is devoted to golf courses.[5]

The flaws in this national formula which had been the basis of the ruling Liberal Democratic party's twenty-year domination became apparent toward the end of the 1960s. First there was the rampant inflation which translated into a 20 percent price rise between February 1973 and February 1974, making inflation in countries such as Britain, the United States, and even Italy pale by comparison.[6] The country's first nationwide general strike since

---

3. Eugene J. Kaplan, *Japan: The Government-Business Relationship* (Washington, D.C.: U.S. Government Printing Office, 1972), pp. 70–71.

4. *New York Times*, 11 June 1973.

5. "Environmental Newsletter," *World*, 17 July 1973, p. 38.

6. *Washington Post*, 7 April 1973. By comparison, between February 1973 and February 1974, the United States experienced a 10 percent

the war brought public services to a standstill in April 1974, as workers demanded wage increases to match spiraling prices. "Prosperity" begins to lose its magic when it is eaten up in inflation. An inadequate social infrastructure meant insufficient sewerage in cities, severe housing shortages, and inadequate assistance for the growing number of Japanese who are living into their sixties and seventies at a time when nuclear families along with cramped urban housing are making the traditional obligation of children to care for older parents untenable. This has been poignantly portrayed in the films of the celebrated director Ozu.[7] Less than 15 percent of Japanese homes are connected to sewer facilities.[8] More than 25 percent of Japanese own automobiles, but highway construction has lagged and roads are increasingly clogged with excessive traffic.

An additional flaw in the formula has been overseas trade and investment's tendency to increase national vulnerability. Japan is a nation with a small territory in relation to its population and with few natural resources. As its economic boom grew in the postwar years, Japan looked overseas for places to invest and expand. Between 1965 and 1972 alone, the value of Japanese overseas

increase; between January 1973 and January 1974 Britain had a 12 percent increase; between December 1972 and December 1973 Italy witnessed a 12.6 percent rise. In Japan's inflationary bracket were India and Israel. Exceeding Japan were South Vietnam and Chile.

7. For a report on the politics of old age in Japan, see David W. Plath, " 'Ecstasy Years'—Old Age in Japan," *Pacific Affairs* 46, no. 3 (Fall 1973): 426–29. Also, Don Oberdorfer, "The Japanese Discover Old Age," *Washington Post*, 4 February 1973.

8. Kaplan, *Japan*, p. 5. In Tokyo with its population exceeding 10 million the diffusion rate of sewage in urban districts is less than 50%, while the national average at the end of March 1971 was only 22.8%. Only an estimated .9 million of the total population of more than 100 million have access to flush-toilet facilities. More than 48% of Tokyo residents surveyed reported to be "extremely dissatisfied" with their housing conditions. Finally, "per capita park space in Tokyo is 1.8 and 2.2 in Osaka, compared with London's 11.4% and New York's 18%." Koji Nakamura, "A Japanese Myth," *Far Eastern Economic Review*, 17 June 1972, pp. 37–38.

private investment grew from $949 million to $6.773 billion.[9] The problems created by this overseas expansion came to a head in the winter of 1973–74. The Arab oil boycott sensitized Japanese policy makers to the danger of dependence on external resources for prosperity. When Prime Minister Kakuei Tanaka took a ten-day tour throughout Southeast Asia he was greeted with protests decrying the creeping Japanese economic hegemony in the area. Among the complaints heard in Malaysia, Thailand, and Indonesia was that, as Japanese voters were growing pollution-conscious, Japanese businessmen were exporting their polluting industries to developing countries that were seemingly too economically needy to worry about environmental hazards. The Japanese are traditionally sensitive to their international image and were shocked by the often violent outbursts that Tanaka's visit provoked, accompanied as they were by symbolic burnings of Hitashi electric fans and Honda motorbikes.[10]

In reaction the Japanese went through a public soul-searching. The foreign minister told the national Diet (legislature) that Japan would have to concentrate more aid on educational and agricultural programs in developing countries. Prime Minister Tanaka then lectured the Diet members and thousands of television viewers: "One hundred million Japanese, in a homogeneous race, have been able to concentrate their energies totally upon restoring and constructing our nation without being hampered by racial conflicts or disputes over religion or

---

9. Bureau of Public Affairs, U.S. Department of State, *Special Report: Japan's Overseas Private Investment—Growth and Change* (Washington, D.C.: Department of State, 23 October 1973), p. 1.

10. American investment in a country such as Indonesia was actually larger than the Japanese investment, but American dollars went toward natural resource exploitation in typically remote jungle areas while Japanese yen were more often backing visible consumer items with easily identifiable Japanese labels.

language." But he went on to warn against the insularity and arrogance that these very advantages could produce. Today that insular posture "is not only unwarranted internationally but also might very well cause troubles." [11] The Japanese would have to learn other languages, be less cliquish when in groups abroad, and perhaps even hesitate before setting up polluting industries in Taiwan or Indonesia.

The deputy prime minister and director-general of the government's environmental agency instructed his agency to work out principles by which Japanese enterprises could protect the environment in developing countries where they were operating, in light of the agency's finding that 106 Japanese firms in Thailand alone were apt to cause pollution.[12] Only nine months earlier, however, the powerful business organization Keidanren sanctioned the sharing of pollution burden in a study entitled "Environmental Pollution and Japanese Industry." The Keidanren report asserted, "There is a physical limit for this country in pursuing the past pattern of economic growth. Many industries are destined to seek sites for their factories in foreign countries with more elbow room from an environmental standpoint." And in an address to the organization the Keidanren president acknowledged that "there is already a rising trend among Japanese petroleum companies to build new refineries overseas due to pressures on the environmental front." [13]

Each of these three issues—inflation, inadequate social infrastructure, and overseas vulnerability and hostility—undermined Japanese confidence in material prosperity and rising GNP, the bedrock of "Japan, Inc." Each issue reinforced public intolerance of pollution and the

11. Quoted in the *New York Times*, 22 January 1974.
12. "Economic Cooperation and Trade: Japan," *Asian Research Bulletin*, 28 February 1974, pp. 250–57.
13. Robert Whymant, "Sharing the Burden of Pollution," *Guardian* (Manchester) 108, no. 27 (30 June 1973).

LDP-business alliance which seemed to sanction pollution as an inevitable cost of economic growth and prosperity. Thus pollution issues gained force because they arose at a time when other issues were prompting a fundamental reassessment of the national goals. In Britain and the United States the environmental issue also raises sharp questions concerning national objectives, but in neither does it challenge one political party apart from the others. In the Soviet Union and Japan the parties ruling when the environmental issue has arisen are parties that are so firmly identified with economic growth that the issue becomes far more party-focused than in either Britain or the United States.

## ISSUE SALIENCY IN JAPAN

Of the four countries investigated here, Japan has given the environmental issue its greatest political saliency. The Japanese government's concern for the environment was manifested initially in 1956 when the Ministry of Health and Welfare proposed legislation called the "Living-Environment and Pollution Control Standards Law." It was widely discussed but ultimately the LDP–cabinet leadership prevented its reaching the Diet floor and it died. The 1950s were a period of heady economic growth, recovery from World War II's destruction and humiliation, consolidation of the conservative party–business alliance dominating Japanese politics. Pollution was scarcely an issue in the public or elite mind. In 1958 a relatively weak Water Quality Conservation Law was passed. Still, the only government people activated by the environmental issue were officers in the Ministry of Health, and they were easily overshadowed by politically influential counterparts in the Ministry of Finance and Ministry of Trade and Industry. Health Ministry officials had to get a grant from the United States to carry on research on fish poisoning.[14]

It was only in the late 1960s that other officials and interest groups gave the issue prominence. The basic document undergirding the Japanese governmental response to environmental disruption is the 1967 Basic Law for Environmental Pollution Control. In 1969 the government enacted a Relief for Victims of Environmental Pollution Law, a bill without precedent in the United States, which provides for the payment of compensation by offenders for medical expenses of pollution-related ailments. This law set the stage for a series of celebrated court cases. The basic law was amended in 1970 during what Japanese have since referred to as the "Pollution Diet." The earlier requirement that antipollution policy be "harmonized" with national economic growth was deleted. Amendments defined pollution as a crime, a charge heatedly opposed by the government's usual allies, big business. Still the new legislation was weakened by its failure to include public gas and electric utilities among enterprises subject to local control.[15]

Stiffening of the laws on paper has not been matched by energetic enforcement. In the United States the gap between policy formulation and enforcement reflects the dual roles of agencies and the fragmentation of power; in the Soviet Union it is due to the industrial managers being entrenched in powerful central ministries. In Japan the gap between policy formulation and enforcement derives from the long-standing alliance between business and the Liberal Democratic party which has ruled Japan since the mid-1950s. In all three countries, however, the gap is maintained as well by ideological commitment to economic growth.

---

14. "Japan: Limits to Growth," *Environment* 15, no. 10 (December 1973): 10.
15. *New York Times*, 11 December 1970.

The bureaucratic manifestation of the Japanese government's environmental policy was the 1971 establishment of the Environmental Agency. It was "designed to coordinate various administrative measures that had been previously undertaken by different Government offices." [16] Actually, the agency has been severely limited in overseeing the implementation of the Basic Law and the eighteen additional laws that were passed in 1970–71. In some cases, only prefectural governors may directly enforce the laws; in other instances, such as laws relating to aircraft noise, the agency has no authority. In still other instances, the agency must hold back or face disciplining by the prime minister.[17]

Between 1971 and 1973, however, four major court decisions resulted in substantial financial penalties against industrial polluters and sharpened the teeth of the government's environmental laws. In the first of the four cases, the Showa Denko Company was found liable for causing "Minamata" disease—from mercury poisoning of fish—in northwestern Japan and was ordered to pay $810,000 in damages to 330 victims. The second court decision came in July 1972, when six petrochemical companies in Yokkaichi, in central Japan, were ordered to pay $285,000 for air pollution causing what came to be known as "Yokkaichi asthma." The government cited 1,054 victims, of whom 76 died. The third celebrated court decision against industrial polluters was handed down in August 1972, against the Mitsui Mining and Smelting Company. A district court in a region west of Tokyo ordered the company to pay $480,000 for dumping cadmium waste into water that was used for rice-paddy irrigation. The waste resulted in what was termed *itai-itai*

---

16. Japan Information Service, *Japan Report* 18, no. 13 (1 July 1972): 2.
17. Kelley et al., "Politics of the Environment: The USA, USSR, and Japan," pp. 22–23.

illness—roughly translated "ouch-ouch"—in which the
victim's skin and bones become painful to the touch. Of
the 265 victims officially cited, 47 died from the disease.[18]
The case that did most to arouse public opinion and
alert Japanese politicians to the gravity of the environmental threat was the fourth case, taken to court by local
plaintiffs in 1969 and finally decided in March 1973.
Mercury-poisoning victims in the small fishing village of
Minamata brought suit against the large chemical firm
Chisso Corporation. "Minamata" disease is named after
the village. The district court on the southwestern island
of Kyushu ordered the Chisso Corporation to pay $3.6
million to 138 individuals representing 30 families. But
the court rulings, while setting crucial precedents, did not
diffuse the environmental issue or guarantee the safety of
Japan's environment. One Minamata disease victim was
quoted as saying: "We want no money. To hell with the
compensation. We want Chisso's management to drink
the waste from their own factory and to know the same
agony." [19]

The court suits suggested the political changes that
the pollution issue might effect in the Japanese political
system as a whole. Three possible trends of far-reaching
significance can be projected from these cases. First, they
represented a public retreat from the post-World War II
Japanese commitment to industrial growth as the formula
for national happiness and international revival. Second,
the cases involved more energetic and effective local level
politics than had been witnessed in Japan. Third, the
cases represented public policy formation outside the
headquarters of government bureaucrats and the ruling
Liberal Democratic party and in direct conflict with the
most powerful interest group and the LDP's staunchest
ally, big business. These trends represent critical steps
away from the conventional model of Japanese politics.

18. *New York Times*, 21 March 1973.
19. *New York Times*, 21 March 1973.

## CULTURE AND ISSUE CREATION

The issueness of pollution depends not only on the public's level of concern, but also on its willingness to press demands. In the United States there is cultural support for organizing and pressing citizen interests on authorities at all levels, though citizen resources are often ineffectual when matched against certain economic interests. In the Soviet Union, citizen mobilization and demand making are hindered by severe shortages of those resources—media outlets, money, official status, electoral platforms—that give citizen activism its political significance. Japan has had political parties, interest groups, and vigorous media for more than a half-century. The American Occupation forces tried to resuscitate these forces, though often under the mistaken assumption that they were nonexistent in prewar years. But the Japanese citizenry has not been comfortable in confrontation. It has preferred patron-client relations, consensus politics, and polite petition over the rougher and mathematically designed style of participation typical of American politics.

Consequently, though the Japanese were becoming increasingly conscious of the social costs of unrestrained industrial growth, most were reluctant to take strong and direct action against those whom they perceived to be the offenders. Initially they preferred to employ polite phrases and mild requests, putting their communications to governmental and industrial officials in terms of appeals, rather than demands or electoral threats. Deference, paternalism, and respect for those in authority dampened forthright action. This inclination is expressed in the Japanese proverb, *Nagai mono ni wa makerero*, which translates, "Let's wrap ourselves up in something long," or more accurately, "Yield to the powerful." [20]

The fact that many people injured by pollution depended on the local offending factory for their jobs also inhibited the transformation of issue consciousness into issue action. Unions were of little help, since they tended to be company unions. The "company town" phenomenon, criticized by generations of muckrakers in the United States, is no less a political factor in Japan.

Building national issues out of local issues is made difficult in Japan because most resources are concentrated in Tokyo. The lack of active political party branches at the grass-roots level reflects this centralization and deprives pollution issues of another possible vehicle for expression at the local level, where they usually are first salient.

But recent change within the Japanese political culture could facilitate issue creation. One study of political socialization of contemporary Japanese teen-agers found that younger Japanese now express more support for popular participation and participatory institutions than do their parents. Likewise, the youths' political culture is moving away gradually from deference to authority.[21] Pollution politics itself may be modifying the traditional Japanese political culture. In other words, a political issue is not simply the passive product to cultural forces, but can be the generator of those forces as well.

Bradford Simcock draws such a conclusion from his investigation of the emergence of the pollution issue in Japan.[22] His cases concern Minamata, where petrochemical plants' dumping of waste into a harbor poisoned fish

---

20. Michael Reich and Eleanor G. Huddle, "Pollution and Social Response," *Area Development in Japan*, no. 7 (1973): 37.

21. Joseph A. Massey, "Political Socialization in a New Democracy: Emerging Patterns of Political Culture in Japan" (Ph.D. dissertation, Yale University, 1973).

22. Bradford L. Simcock, "Environmental Pollution and Citizens' Movements," *Area Development in Japan*, no. 5 (1972): 13–22.

that were then caught by the local fishing fleets and spread mercury poisoning to local consumers for whom fish is a dietary staple; and Toyama, where rice crops were contaminated by industrial cadmium waste. In "Phase 1" of the issue movement citizens in these cases appealed for compensation for damage already done; in "Phase 2" Japanese went beyond traditional appeals to organize for the sake of long-term prevention of such pollution tragedies. In addition, Simcock notes that in the early phase of both cases participants and the issue definitions were localized, but in the later stage Japanese from outside the districts came in to offer professional expertise or moral support. The issue thus expanded from health of a fishing village or farming community to a question of the ruling Liberal Democrats' national priorities and big business's public accountability. The company-town context in which both pollution tragedies occurred stifled protest initially and reinforced the tendencies to adhere to patterns of paternalistic deference. But the very domination of the town by companies made resort to external allies all the more necessary for the injured claimants.

In Minamata the obstacles to moving from Phase 1 to Phase 2 were almost insuperable. There was little local autonomy or civic spirit; the government-allied private chemical firms had resources that modest fishermen could not hope to equal. Only when the national press began to publicize the mercury-poisoning incident and the Ministry of Health began collecting names of victims who could collect damages did other sectors of Japan focus on Minamata and its national significance. It was one of the leading national newspapers, *Yomiuri Shimbun*, with a circulation of 5 million, which at the height of the Minamata case invited Ralph Nader to make an environmental tour of Japan.[23] This too served to broaden

---

23. Donald Kirk, "Students in the Elementary School Grow Up Suffering from Asthma . . ." *New York Times Magazine*, 26 March

the issue and to mobilize Japan's frequently ineffective leftist parties to take an interest in the local welfare of Minamata and Toyama. Simcock concludes that such Phase 2 issue development in Japan is not widespread. But as various issues converge to raise profound doubts about the priority given to economic growth, the authority of those leaders or institutions that have been most identified with that single-minded policy is also losing its traditional deference. Organization for the sake of prevention will replace activity for the sake of immediate compensation in those cases where one company or industry does not monopolize a community, where the citizens are less imbued with Japanese rural values, and where national publicity prompts outside groups to become actively involved. Finally, Phase 2 politics is more likely where progressive local politicians see pollution as a weapon for building up their own personal power, thus running for office more on issues and less on the traditional personal loyalty groups. Minamata's mercury poisoning resulted in deaths and national issue-consciousness, but antipollution movements in urban settings such as Mishima-Numazu, where citizen groups prevented the construction of a petrochemical complex, represent a more politically assertive Phase 2 sort of "new politics" in Japan.[24]

## DEVELOPMENT AND THE ENVIRONMENT

As in other industrialized countries, economic growth and urbanization have been distributed unevenly in

---

1972, p. 64. The Japanese press has been accused both of being too antigovernment and too cozy with the regime. The frequent timidity of Japanese newspapers is due largely to journalists' personal intimacy with politicians and bureaucrats as part of the establishment. *New York Times,* 3 December 1973.

24. Simcock, "Environmental Pollution," p. 19.

Japan. Most of the pollution protests have occurred on Japan's most densely populated island, Honshu. Today Japanese government planners are looking more and more to the northern island of Hokkaido. With 21 percent of the nation's land but only 5.3 percent of its population, Hokkaido long has been a backward agricultural region. Now it is being viewed as a possible safety valve. Hokkaido cannot possibly replace Brazil or Indonesia in the Japanese economy, but it may give Honshu some relief from its industrial and population pressures.

The central government's ten-year development plan budgets $25 billion and private investment totaling $30 billion for Hokkaido. By 1980 the manufacturing production value of the island is intended to multiply fivefold so that Hokkaido will match Honshu in industrial capacity.[25]

However, the ambitious Hokkaido plan has met increasing opposition from the newly potent environmental movement. Environmentalists have objected to coalmine, atomic-reactor and oil-refinery sitings on the island. Fishermen, a powerful lobby in Japan, have won withdrawal of one plan for an industrial zone in the Hokkaido port town of Hakodate. Student petitioning has blocked construction of a road through virgin forests.[26]

In response to opposition, the government stressed that stringent pollution controls would be imposed on any new industry. Still, opposition party politicians revealed that government agencies had submitted false environmental estimates on one such development scheme for Hokkaido. The regime also notes that 9 percent of the island will be devoted to national parks.[27]

Strapped by limited land space and minimal natural resources, Japan's development dilemmas are even more severe than those of the United States. Japan does not

25. *New York Times*, 3 May 1974.
26. Ibid.
27. Ibid.

have the markedly uneven distribution of wealth in its
citizenry that makes growth such a necessity in the
United States, and it has been among the world's most
successful practitioners of population control. But Japan
is a vital and innovative country with an upwardly mobile
and consumerist population and an industrial base that
needs raw materials and outlets for its capital. In Japan
the debate between development and environment takes
on more immediate foreign policy significance, for overseas investment is especially critical to the health of the
Japanese economy as it is currently structured. As its
citizens at home become less and less tolerant of pollution, Japanese business executives and government
officials look all the more eagerly to foreign countries for
sites of unwanted industries, at just the time when
environmental consciousness is beginning to gain saliency in those countries as a dimension of nationalism.

## GROUPS AND PROCESSES

The Japanese policy-making structure has been labeled "a kind of unholy trinity." [28] Since the consolidation
of political power under the Conservative party in the
mid-1950s, this "unholy trinity" has included (1) the
Liberal Democratic party (LDP), (2) top-level professional
bureaucrats, and (3) big business. The cohesiveness of
these three elements and the widespread public acceptance of their authority has given Japanese politics its
closed, stable, and purposeful character.

The basic antipollution law of 1967 and stronger bills
passed since then have had to make their way through a

28. Hiroshi Itoh, "Editor's Introduction," in *Japanese Politics: An
Inside View* (Ithaca, N.Y.: Cornell University Press, 1973), p. 3. See also
Shigeo Misawa, "An Outline of the Policy-Making Process in Japan,"
*Japanese Politics: An Inside View*, pp. 12–48.

political process that has permitted little effective policy making outside these three forces. More recently, however, mobilized antipollution movements have begun to make at least some minor modifications in this process. First, they have bolstered local governments' role in Japanese policy making, a role which, despite American occupiers' efforts, has remained attenuated. Second, antipollution movements have undermined the hegemony of the Liberal Democratic party by giving the usually fragmented, ineffectual opposition parties on the Left an issue that has more direct meaning to Japanese voters than their usual ideologically coated platforms. Third, public anxiety over the environmental hazards from unrestrained growth has limited the power, or at least altered the policy priorities, of the Ministry of Trade and Industry and the Ministry of Finance. Finally, there may now be emerging in Japan a political force—organized citizenry equipped with voting power, expertise, and legal standing in the courts—that can act as a counterweight to big business.

Each of these trends is more potential than real at present, though many Japanese reformers hoped that the resignation of Tanaka and the ascendance of Takeo Miki to prime minister in 1974 reflected acceleration of these trends. Currently, efforts to formulate policies protecting the environment still have to work their ways through the councils dominated by the "unholy trinity."

The LDP's 20-year political hegemony has been due to its success in reasserting Japan's influence and autonomy after the wartime defeat and occupation. Its principal power bases have been the business establishment, which supplied funds for LDP candidates, and rural constituents, who have provided reliable though now dwindling votes. LDP's big business connections give the chief opponents of antipollution programs a special access into the decision-making process. For example, during the celebrated "Pollution Diet," the parliamentary session of 1970, big business representatives were exercized over the amendment to the Environment Law that would have made pollution a crime. As originally drafted by the LDP cabinet and its bureaucratic consultants, the bill provided for seven years in jail or a fine of 5 million yen (approximately $16,000) for firms or individuals convicted of industrial polluting; and such responsibility was to be tightly defined. Business spokesmen labeled the bill vicious and unrealistic. The Chamber of Commerce and Industry as well as the Japan Committee for Economic Development, business's main lobbying organizations, went straight to the top executives of the LDP to lodge their protests. Another influential group, the Federation of Economic Organizations, applied pressure within the Diet, arguing "We do not want instant pollution control. . . . It is absurd to institute penal provisions against the crime of pollution when such a dearth of scientific evidence as to what relationship there is between cause and effect exists." [29] The opposition party Dietmen charged business interests with threatening to withhold election funds from the LDP if the regime passed the bill.

Under pressure from their closest ally, the LDP leadership revised the "Crime of Pollution" bill. This was not done in open parliamentary debate, but behind the scenes, as is typical in Japanese policy making. The critical change was cutting out the words "the possibility of" from the key legislative provision making criminally liable "those who have given rise to conditions with the possibility of causing danger to the lives of people in general or their bodies." [30] The Japanese press and public

29. Quoted in Darrell Houston, "Remember When You Could See Mt. Fuji?" (Report of the Alicia Patterson Fund, New York, 13 January 1971), p. 4.
30. Ibid., pp. 4–5.

are accustomed to such collaboration between the LDP and big business, but this blatant surrender under pressure appeared to be particularly offensive, perhaps because the legislation was so directly concerned with the personal welfare of ordinary citizens and because it was linked to an issue which had gained such high political visibility.

Japanese business lobbies remain the single most influential extraparty force shaping LDP policy decisions, but business may be losing some of its postwar invulnerability in the eyes of Japanese citizens. Four years later, in the midst of the Arab oil boycott and accelerating domestic inflation, a Diet committee took the unprecedented step of calling corporate executives before it to berate them publicly for alleged petroleum price manipulations. The businessmen were compelled to offer humble apologies to the public for their practices.[31]

Each of Japan's political parties is led by its Dietmen. But within each party the Dietmen are split into factions based not on ideology or regional ties, but on personal loyalties and obligation to particular party leaders. Since the LDP's ascendancy in 1955, the Diet's real policy-making role has diminished. The majority of bills introduced come from the executive. There is provision for private bills introduced by individual Dietmen, but these are far fewer in number and more peripheral in policy importance. It is unlikely that a Japanese Edmund Muskie could sponsor and shepherd through the legislative process so major a bill as the Air Quality Control Act. Japan's

---

31. *Boston Globe*, 3 March 1974. See also Chitoshi Yanaga, *Big Business in Japanese Politics* (New Haven: Yale University Press, 1968). Further evidence of Japanese businessmen's reassessment of their role was their apparent cuts in funds donated to the LDP after the July 1974 House of Councillors election in which business organizations and the LDP were harshly criticized for their lavish campaign expenditures. See Koji Nakamura, "Cutting Off the Source of Cash," *Far Eastern Economic Review*, 30 August 1974, pp. 25–26.

opposition parties can introduce bills, but holding usually only one-third of the lower-house seats, they cannot hope to pass many. The government's proposals come from either the ministries by way of the cabinet or from the LDP's various party policy committees. It is standard procedure for the party committees and the relevant ministry to work together in research and drafting legislation.[32] Close relations between professional bureaucrats and LDP Dietmen serve to exclude all but the most privileged and well-connected outside actors.

The bureaucrats have the expertise, which the LDP often lacks, and the LDP has the legislative authority that the bureaucracy needs. There are several reasons for this easy working relationship between the two. Many civil servants share the LDP's basic definition of Japan's national goals, especially its need to restore itself to international prominence via nonmilitarist industrial power, a goal that can be achieved only through orderly, centralized, and carefully planned policy moves. Many civil servants and LDP Dietmen also share similar schooling, coming out of the large universities like Tokyo University. Most noteworthy perhaps, is the frequent interchange between bureaucratic and party personnel. In contrast with the United States, but not unlike the Soviet Union, Japan's government bureaucracy serves as a training ground for many future party leaders. All of the post-1955 LDP prime ministers and many of the cabinet ministers have served as bureaucrats before becoming full-time elected politicians.[33]

A younger generation of LDP Dietmen has grown

---

32. One of the most detailed analyses of the intraparty workings of the LDP is provided in Nathaniel B. Thayer's *How the Conservatives Rule Japan* (Princeton, N.J.: Princeton University Press, 1969), especially pp. 207–67.

33. Key Sun Ryang, "Postwar Japanese Political Leadership—A Study of Prime Ministers," *Asian Survey* 13, no. 11 (November 1973): 1015.

restless with what they see to be the closed leadership and lack of issue orientation within the party. The LDP "young Turks" are by no means united in their policy stances, but they share some important common characteristics. They are men in their thirties and forties whose political memories are of the postwar Japanese economic boom, not of the prewar chauvinism and wartime humiliation; they are less psychologically scarred and thus less imbued with caution. They are more at ease with foreigners and have a more direct political style, eschewing their elders' traditional preference for indirection and deliberate vagueness. They are less likely to come from the traditional elite and more likely to have degrees from private institutions rather than from Tokyo University. And they have expressed concern with the LDP's apparent increasing alienation from the average Japanese voter. Among the issues that this rising political generation have been urging the older leadership to take seriously are defense (an army is prohibited under Article 9 of the postwar constitution), inflation, and pollution.[34]

When Kakuei Tanaka succeeded Eisaku Sato as prime minister and LDP leader in July 1972, it seemed at first glance that indeed the mantle had been passed to this new generation. Although Tanaka was older chronologically, he shared with them a nonelitist education, a stylistic directness, a businessman's pragmatism, and a willingness to take risks. At first he rarely listened to his cabinet or top bureaucrats and was dubbed "the computerized bulldozer." His best-selling book called for a radical demographic remodeling of Japan, based on a dispersal of people and industry away from the clogged urban centers to the islands' less developed regions, such as Hokkaido.

---

34. Richard Halloran, "Young Politicians in Japan Chafe at the Old Rigidities," *New York Times*, 12 March 1974. Also *New York Times*, 2 October 1973.

Within twelve months of taking office his ratings in the Japanese opinion polls dropped from their original peak of 65 percent favorable down to a mere 25 percent.[35] His much-heralded decentralization scheme had bogged down in the Diet. Opposition parties won more local elections. Land prices, already soaring, were further inflated by prospects of the Tanaka decentralization plan. Voters living in the less densely populated regions experienced second thoughts about welcoming the industrial overflow from Osaka and Tokyo.[36] There was suspicion that Tanaka's enthusiasm for industrial and urban resettlement had more to do with his own construction business and Ministry of Construction background than with his worry about the national environment. And the younger LDP Dietmen, such as those in the much-publicized Summer Storm Society, stepped up their criticism of the party's leadership's inability to cope with the country's problems. Finally, inflation and pollution seemed to voters to be as severe as when Tanaka came into office. The crisis of confidence reached a climax in the autumn of 1974 with the revelation of Tanaka's dubious financial dealings while serving as minister of construction. Japanese, close followers of American politics, dubbed the scandal "Japan's Watergate." Tanaka was forced to resign, leaving his succession in the hands of senior LDP statesmen. Resisting young Dietmen's call for an open nominating convention, the elders chose a surprise compromise candidate as new party leader and thus prime minister, Takeo Miki.

Optimism upon both Tanaka's and Miki's successions was perhaps overblown and did not take sufficient account of the structural conditions which are hindering any government's ability to deal with Japanese environ-

35. "Japan: It Could Happen Here," *Economist*, 2 June 1973, p. 45.
36. Donald Keene, "Letter from Tokyo," *New York Times Magazine*, 3 March 1974, p. 26.

mental dilemmas. The prime ministership is not equivalent to the American Presidency. A charismatic Japanese prime minister is almost a contradiction in terms. For the role is severely circumscribed by the factionalism within the political parties and the norm of behind-the-scenes consensus building that shapes decision making. In addition, the LDP lack of grass-roots organization inhibits its potential for mobilizing the citizenry behind new programs.

Thus it may be more important to look outside the Liberal Democratic party for the changes that could alter Japanese policy processes. Those changes are occurring. Moreover, they are occurring in large part in response to the public issue awareness concerning pollution. In particular, three political trends have emerged in recent years: (1) the rise of environmental interest groups willing to go beyond traditional deferential petitioning to public mobilization for explicit demand making; (2) the new relevance and autonomous activism on the part of Japan's long quiescent local governments; and (3) the new success of opposition parties in local organizing and winning local elections.

## INTEREST GROUPS AND LOCAL POLITICS

Interest groups in Japan are less clearly differentiated from political parties than in either Britain or the United States, though not as firmly coopted as in the USSR.[37] Business organizations enjoy close connections with various LDP factions, while labor unions are the principal sponsors of leftist party candidates. Nevertheless, antipollution groups are more autonomous from parties. The Japanese Communist party has been the most successful

---

37. For example, see William E. Steslicke, *Doctors in Politics* (New York: Praeger, 1973), p. 233.

in absorbing new environmentalist groups into its party structure, partly because the JCP has been the quickest to see the need for building party roots at the local level rather than relying, as the LDP does, on national level organization exclusively.

Antipollution groups demonstrated their impact on policy makers by halting the construction of new power plants which had the Japan Atomic Energy Commission's as well as MITI's backing.[38] Similarly, ad hoc alliances of farmers and radical students stalled construction of a large Tokyo jet airport in what came to be termed the "battle of Narita," after the protestors were evacuated by riot police.[39] Dr. Jun Ui, an urban and environmental specialist at Tokyo University, told an international conference on the environment that Japan's traditional legal theory and the reluctance of the current regime to act, made such demonstrations necessary: "I don't say that only violence works, but so far we haven't had the legal instruments as such, and what we've been able to do we've done by the political instrument; namely, the demonstration." [40]

It has been on the local level that autonomous antipollution groups have had the greatest impact on the policy process. They have compelled local decision makers to deal with newly assertive campaigners, where traditionally they have had to cope with deferential petitioners. Citizens have employed not only campaigns, but also media-attracting demonstrations and law suits. Their activities have prompted a slow movement toward a less personalist, more issue-oriented form of electoral behavior in Japan.[41]

38. "World Environment Newsletter," *World*, 31 July 1973.

39. *New York Times*, 6 March 1971, 17 September 1972.

40. Quoted in Fred Shapiro's report on the first interdisciplinary conference on environmental pollution, Tokyo, May 1970: "Our Far-flung Correspondents: E.D.," *New Yorker*, 23 May 1970, p. 103.

41. Margaret A. McKean, "The Potential for Grass-roots Democracy in Post-industrial Japan: The Anti-pollution Movement" (Ph.D.

Most successful in forestalling construction of polluting industries or in gaining better garbage facilities have been groups that combine traditional methods and appeals with more modern interest group tactics. Effective also have been Japanese groups which concentrate on the local dimensions of problems since a local focus can enable conservatives and progressives to work together. This formula, however, has meant that ideology has had to be soft-pedaled. Consequently, although an antipollution movement may inspire its followers to vote for the opposition Socialist or Communist parties in a local election, those votes may contain minimal ideological content and be aimed principally at frightening the arrogant LDP. Furthermore, the antipollution groups suffer from conflicts with one another. This has been especially true in Tokyo, where antipollution groups organized by neighborhood wards have been primarily concerned about keeping new garbage incinerators out of their own environs, rather than with hammering out a city-wide garbage policy that will treat all wards justly.[42]

The Socialist and Communist parties (JSP and JCP) have been far quicker than the LDP to adapt to these upsurges of local political assertiveness. The two parties, either in tentative alliances or acting separately, succeeded by mid-1973 in winning important mayoralty elections in such cities as Tokyo, Osaka, Kyoto, Kobe, Yokohama, and Nagoye. By focusing their party energies on the municipal and prefecture (equivalent to county) level, the opposition has served to increase the political importance of local government. On the other hand, winning the governorship of Tokyo prefecture is not the same as winning the governorship of New York State, for local autonomy is very limited in the Japanese system.

dissertation, University of California, Berkeley, 1974). The following discussion draws on chapters 7 and 8.

42. Ibid.

Recent environmental legislation has delegated to the prefectural governors wider powers in the battle against pollution, but they remain wholly dependent on the central government for the budgetary resources with which to exercise that new authority. A city in Japan may not make laws or even ordinances; prefectural governors may make ordinances but these can be easily overturned by the central government. The most politically astute of the opposition prefectural and municipal officials have used this limited authority where possible but have also taken advantage of their new visibility in the legal structure to publicize local issues and demonstrate the inadequacy of the LDP's central regime in coping with local needs.[43]

After decades of being unable to break through the one-third block, the opposition parties in 1970 decided to forgo some of their ideological commitments in order to present a united opposition candidate in local elections. The tactic became known as the "Takatsuki Formula." In Takatsuki, a town of 220,000, the Japanese Socialist party, the Japanese Democratic Socialist party (a splinter off the older JSP) and the *Komeito*, or "Clean Government," party banded together and won. The Communists at that time refused to soft-pedal their Marxist ideology and so stayed outside the alliance. More recently the JCP, in a campaign to strengthen its position in Japanese politics, has joined in similar local alliances. The principal issue that has served as a substitute for ideology has been pollution.[44]

---

43. See, for example, the *Tokyo Municipal News*, a monthly journal published by the socialist government of Tokyo, available in English from the External Affairs and Tourist Division, Bureau of General Affairs, Tokyo Metropolitan Government, Marunouchi, 3-5-1, Chiyoda Ward, Tokyo, Japan. The newsletter gives regular coverage of environmental conditions and policies.

44. Michael Lyons, "Environmental Pollution and Party Politics in Japan" (Paper, Miami University, 1970), pp. 10–13.

The Socialist governor of Tokyo, Ryokichi Minobe, elected in 1967, was one of the earliest beneficiaries of this new leftist party strategy. A politician is not only the beneficiary, but a captive of an electoral strategy, so Minobe has played down his party role and instead run as an issue man. Pollution has been Minobe's issue. He welcomes the appellation "Mr. Anti-pollution" and advocates stronger local government, contrary to the orthodox socialist preference for centralization. He takes credit for writing tougher antipollution ordinances than the central government, for instance in the matter of industrial smokestack specifications, defying Japanese tradition in which central policy is stricter than local. The Environmental Agency officially permitted local authorities to make stiff restrictions on industrial wastewater in cities over 500,000.[45] He argues that local governments are closer to the people and can best serve their immediate interests. In that same vein, Minobe has contended that his administration has done more to encourage popular participation in city administration.[46] This image of open policy making has encouraged antipollution mobilization but has made Minobe the target of conflicting ward groups, none of whom want waste plants.[47]

In Tokyo the Communist party has supported socialist Minobe, though it has run its own candidates for the municipal assembly, in which the LDP continues to hold a majority.[48] In Kyoto the JCP has played a more dominant

---

45. McKean, "Potential for Grass-roots Democracy," chap. 7, p. 26.

46. Ryokichi Minobe, "Improving the Environment in Tokyo: Pollution of the Atmosphere," *International Union of Local Authorities Newsletter*, January 1974, pp. 1–2. See also David K. Willis, "Tokyo's Mr. Anti-pollution," *Christian Science Monitor*, 18 September 1970.

47. McKean, "Potential for Grass-roots Democracy," chap. 7, pp. 7–20.

48. For an analysis of the 1973 Tokyo election in which the LDP scarcely maintained its municipal assembly majority, see Tosh Lee, "Tokyo Metropolitan Assembly Election—1973," *Asian Survey* 14, no. 5 (May 1974): 478–88.

role in the opposition alliance that controls the prefectural government. Kyoto is perhaps the JCP's most successful experiment with its new "soft" image, which emphasizes grass-roots organization and local issues over its former reliance on abstract ideology and labor-union support.[49]

The seventy-seven-year-old prefectural governor, Torazo Ninagawa, eschews partisan designation, preferring to label himself a "progressive." He relies on a political formula which includes: "vocal disdain for the conservative central goverment in Tokyo; resistance to industrial penetration of the old imperial capital, benefits for small businessmen and fealty to the Kyoto traditions." [50] In the early years of the Ninagawa regime the Socialists were the dominant partner in the coalition, but the JCP has demonstrated increasing electoral strength locally and has become the second most powerful party in Kyoto after the LDP, which, as in Tokyo continues to win majorities in the local assembly. The Kyoto branch of the JCP has been assiduous in building support among laborers, municipal employees and small businessmen. The JCP has not introduced radical programs, for the antipollution issue in a traditional city such as Kyoto is a resistance to change. The governor has said, "[Kyoto] has its own history of 1,000 years, and people here have this city in their blood. Though we have hamburger stands in Kyoto these days, this place is not as westernized as Tokyo. These are Kyotoises, and they will remain so." [51]

Compared to its own past, Japan's policy process in the field of environmental affairs appears more activist, issue-oriented, and decentralized today. But in a cross-

---

49. George O. Totten, "The People's Parliamentary Path of the Japanese Communist Party: Part II—Local Level Tactics," *Pacific Affairs* 46, no. 3 (Fall 1973): 384–406.

50. Don Oberdorfer, "It's a Joyful Spring for the Tiger of Kyoto," *Washington Post*, 17 March 1974.

51. Ibid.

national comparison Japanese environmental decision making appears ineffective. For instance, Akira Nakamura's comparative investigation of air pollution policy making in Los Angeles and Osaka reaches a conclusion that might surprise many smog-oppressed Californians.[52] He found that Los Angeles has continually outdistanced the equally smoggy city of Osaka in taking authoritative action to limit air pollution. The factors he cites as the chief causes for the discrepancy bear on the larger question of the Japanese political system's capacity to cope with the new demands for environmental control. First, the Osaka government has no financial resources and is totally dependent on the budgetary allocation politics controlled by the Liberal Democrats in the central government. Los Angeles is plagued with insufficient tax revenues and increasing reliance on state and federal funds, but still has more financial discretion locally than does Osaka. Second, Osaka confronts not one but a vast number of central authorities which have responsibility for air quality. Because they are so dependent on the center for their funds, "local governments produce one rule after another in order to ingratiate themselves to the central government, although they are unable to register substantial progress in the alleviation of smog." [53]

Like the prefectural governors of Tokyo and Kyoto, the governor of Osaka came into office on the coalition support of opposition parties in 1970. As a Socialist, the governor pledged to end the pro-business stance of his conservative predecessors. In the realm of air pollution control this meant a revitalized plan for clean air, conferences of health officials, and establishment of more stringent air-quality ordinances. But, despite these actions, the air conditions over Osaka have not markedly improved during the first two years of Socialist party rule.

52. Nakamura, "Japanese Myth."
53. Ibid., pp. 234–35.

The reason lies in the extremely limited powers held by Japanese governors and mayors. Added to this is the governor's role as a Socialist, which makes his access and influence in the central ministries all the more restricted. On the other hand, the mayor of Osaka has been somewhat more effective in dealing with the central government in environmental matters. While the mayor was also a Socialist-backed candidate, he had worked as a civil servant in Tokyo, in both the Ministry of Foreign Affairs and the Ministry of Labor. Furthermore, unlike the governor, the mayor had had years of experience in city government, first as a liaison officer from Labor and then as deputy mayor.[54] In a political system where personal contacts count so heavily, these background factors are crucial for effective political decision making.

Nakamura found that Los Angeles politicians and technocrats couldn't ignore the problems of air pollution because the contest between the environmentalists and their opponents wouldn't die. By contrast, in Osaka there was less open controversy. Osaka business interests were strong, but they rarely had to battle openly with well-organized environmentalist lobbies. Thus, while Los Angeles's air might be cleaner today if American environmentalists were better equipped in the political arena, even unequal contests had a beneficial effect: air quality remains a lively political issue. The groups in Osaka, by contrast, are too weak even to keep the issue in the forefront, much less to win legislative and bureaucratic battles.

Nakamura's study modifies the projections of observers who note the emergence of activist citizen groups in Japanese politics. Nevertheless, he and they agree in pointing to ideologization as a major weakness in many Japanese environmental groups: "Although these groups

54. Ibid., pp. 237–38.

have held mass rallies and public discussions, they have lacked specific points of criticism or pragmatic solutions as an alternative to the present control system. Instead these groups only pointed an accusing finger and charged that pollution was a negative result of the political and economic system of Japan." [55]

## BUREAUCRATIC POLITICS

Bureaucrats have played critical policy-making roles in Japan ever since the Meiji oligarchs set Japan on its modernizing course in the 1860s. They have been not only administrators but also designers of the country's long-range development. Their influence has been felt especially in economic policy making. Bureaucrats have an autonomy and high social status that would surprise Americans. They work closely with the ruling LDP, though they are not as intimately intertwined organizationally with the dominant party as are Soviet government officials with the CPSU. Their status and educational background compares favorably with those of British civil servants, though the latter might be more shy about admitting such direct policy influence.

Japanese bureaucrats are an elite based on talent. They are recruited from the top graduates of the nation's public universities, where admission is intensely competitive. Sixty-three percent of Japan's top administrative officials in the Ministry of Trade and Industry, for example, are graduates of Tokyo University, as are 73 percent of those in the Ministry of Local Government and 62 percent of those in the Ministry of Finance.[56] Retirement is relatively early for top-ranking civil servants. At the

---

55. Ibid., p. 196.
56. Akira Kubota, *Higher Civil Servants in Post War Japan* (Princeton, N.J.: Princeton University Press, 1969), pp. 140–46.

higher levels the average retirement age is 50.9 years, while in the economic ministries it comes even earlier.[57] But "retirement" is short-lived, since it is common for many civil servants to go into politics, running for the Diet, usually under the LDP banner. The majority of Japanese retired civil servants, however, take up high-paying posts in private industry or government-run corporations. Those ministries offering top officials optimal chances for rewarded mobility are the ministries with closest ties to industry, especially the influential ministries of Finance, Industry and Trade, Forestry, and Transportation.[58]

Japanese bureaucrats share many cultural inclinations with their fellow countrymen in the Diet, political parties, and interest groups: hierarchical notions of social relations, a tendency to identify with groups rather than as free-floating individuals, avoidance of open confrontations. Hierarchy and social deference serve to create barriers between Japanese officials and ordinary citizens, which in turn limits the latter's access to administrative offices where so many crucial decisions are made. Although on the surface Japanese bureaucrats appear highly legalistic, dedicated to the law's fine print, they are in fact primarily concerned with lessening risk and ensuring anonymity; resorts to legalism simply provide a welcomed retreat from a difficult situation.[59] Like the Dietmen, they are anxious that decisions be supported by consensus and that no action seems to be taken on one person's or one bureau's initiative. This leads to what critics claim to be endless consultations and vague lines of responsibility. Japanese refer to their bureaucratic decision-making system as *tsumiagishiki*, the "piling up

57. Ibid.

58. Chitoshi Yanaga, *Big Business in Japanese Politics* (New Haven: Yale University Press, 1968), p. 98.

59. Nobutake Ike, *Japanese Politics: Patron-Client Democracy* (2nd ed.; New York: Alfred A. Knopf, 1972), pp. 70–71.

method," which progresses "step by step through appropriate channels until the completed decision reaches the vice minister at the top." [60]

Japan's bureaucrats also share in the nation's orientation toward group-defined identities. They tend to think of themselves first as members of their own departments and second as officials of a single national administration. This group identification is stronger than the mere self-interest which prompts an American Forestry Service official to look at a policy choice in terms of the service's welfare. Just as Japanese industrial and office workers rarely move from company to company, so Japanese bureaucrats typically stay in the ministry they join after the university and see their chances for promotion and postretirement largely in terms of that particular ministry. Such "groupism" leads to sectionalism and empire building. It makes proposals for administrative mergers or cooperation especially hard to implement.

The departments enjoying the greatest political prestige and influence are the Ministry of Finance and the Ministry of Trade and Industry. Japanese prime ministers and cabinet members are bred here. Officials in these ministries have spent generations building up close working relations with the nation's major manufacturers and large trading firms. It is not simply cultural tendencies that account for these intimate civil service–business networks. There are structural explanations as well. For instance, a recent study of Japanese decision-making processes revealed that the decline of the parliament in policy formation, the increasing use by government of advisory bodies which include business executives— many of whom are ex-bureaucrats—the growing reliance on administrative directives in place of parliamentary legislation and, finally, the ruling party's dependency on

---

60. Yanaga, *Big Business*, p. 99.

bureaucratic agencies for electoral campaign proposals and legislative drafts—all are not only fostering bureaucratization of politics but also ensuring close cooperation between top-level bureaucrats and major business organizations.[61]

The Ministry of Trade and Industry has been most visible in its resistance to the imposition of environmental controls. It has feared that strict controls and stiff penalties would injure Japan's economy and thus undermine its international security, which is based on GNP rather than guns. In most instances, MITI apparently "wishfully neglects the issue" in hopes that it will go away.[62] But there have been a few cases of blatant interministerial conflict over pollution issues. During the outcry over the Minamata fish poisoning MITI opposed the Ministry of Welfare and Health's guidelines which directed that all mercury dumping cease by September 1974. MITI intervened and the deadline for the chemical plants was postponed until a year later.[63] Under usual circumstances the Health Ministry would have been a poor match for MITI. But in the Minamata controversy officials had the backing of a mobilized local citizenry and an attentive national press. In Japan it is less common than in the United States for an administrative agency to take part in this sort of constituency mobilization as a strategy for the sake of bolstering its own negotiating position within government.

MITI also took part in the legislative debate over the polution-as-crime bill introduced during the 1970 "Pollution Diet." Ministerial bureaucrats were appalled at the wording of the bill, which would have made industrial

61. T. J. Pempel, 'The Bureaucratization of Policymaking in Postwar Japan," *American Political Science Review* (November 1974).

62. Margaret McKean, personal communication with the author, 26 February 1974.

63. "Japan's Pollution: Ouchi-ouchi," *Economist*, 25 August 1973, pp. 77–78.

firms liable for injuries derived from polluting practices. MITI officials were active during the behind-closed-doors rewrite sessions and pressuring the LDP Policy Council members to weaken the bill so as to allay the fears of the party's business supporters while still responding to widespread public anxiety over pollution. The compromise version of the bill which emerged from these sessions was partly the handiwork of MITI bureaucrats.

In any political system, as the public's priorities change so too does the role of particular bureaucratic agencies. If the Japanese leadership and voters downgrade economic growth as a national priority and assign more weight to such goals as social justice or environmental health, MITI's officials will have to make their own adjustments. They could take the form of encouraging Japan's manufacturers to capture new markets with antipollution technology that is eminently exportable. More drastically, MITI's role could be redefined from outside.

In 1973 the Tanaka government proposed legislation that would curb MITI, partly as a means of clipping the wings of Tanaka's own cabinet rivals and partly as a way of capitalizing on rising antiindustry sentiments at a time when the prime minister's own popularity was dropping. Tanaka's far-reaching reorganization of MITI might have stemmed, too, from his own firsthand knowledge of MITI's autonomous power gained while he headed the ministry. The prime minister's proposal would take away from MITI one of its key functions, industry-wide economic planning, transferring it to the Finance Ministry and leaving MITI to work out mere details. MITI also was ordered to create a new bureau for the environment that would exercise authority over the existing pollution study bureau and the department in charge of recycling, thus strengthening MITI's commitment to environmental control from within rather than depending wholly on external monitoring. The result overall would be that MITI's policy role would be truncated, especially in the realm of heavy industry planning. The ministry does have new responsibilities for energy policy and for liberalizing the nation's trade laws, but it is not the same MITI as before. On the other hand, MITI is no "helpless giant" that Japanese environmentalists can ignore. For it still is headed by leaders of the LDP's most powerful intraparty factions. And only a month after Tanaka's proposals were introduced, MITI was able to obtain pollution compromises from the Health Ministry.[64]

An ambitious, bright Tokyo University graduate would still be more likely to seek a government career in MITI than in the fledgling Environmental Agency, created in 1971 and located within the prime minister's office. Prior to its creation, environmental affairs were scattered in several ministries and had minimal budgetary resources. The first director of the new agency was a medical doctor and outspoken advocate of environmental control, Buichi Oishi. He clashed with the LDP leadership and held up several construction projects opposed by conservationists. Dr. Oishi saw himself as a spokesman for antipollution groups who otherwise had little access to the inner circles of Japanese politics. Under his directorship the agency issued an official white paper calling for a switch from what it termed passive policies to more positive control and management of the environment. The report went on to cite the human and physical costs of economic growth, estimating that they totaled $5 billion per year in Japan.[65]

---

64. "MITI's Been Cut Down to Size," *Economist*, 28 July 1973, pp. 71–72. Later in 1973 MITI issued its first energy "white paper" in which it stressed that any new energy production strategies would have to be in line with environmental preservation. "Japan's First White Paper on Energy Stresses Conservation, Cooperation," *Japan Report* 19, no. 24 (16 December 1973): 1–3.

65. *New York Times*, 29 May 1972. See also "Japan Taking Firm

Dr. Oishi's assertiveness eventually prompted his ouster by the prime minister in favor of an administrator who would be less offensive to the LDP's business allies. His successor as head of the Environmental Agency was Osanori Koyama, an administrator who was formerly minister of construction and had close ties with the conservative wing of the LDP.[66] The third head of the agency was even more a product of the political mainstream. Takeo Miki, in his late sixties and leader of one of the chief LDP intraparty Diet factions, was one of the party senior spokesmen whom Tanaka had to mollify in order to maintain his hold on the prime ministership. Head of the Environmental Agency was not considered as much of a political plum as heads of either the Finance Ministry or MITI, both of which went to LDP politicians leading even more powerful factions. Nevertheless, with the selection of Miki the Environmental Agency director was elevated to the status of state minister, thus giving the agency direct access to the cabinet. Miki was also made deputy premier.[67] However, Miki's selection to succeed Tanaka as LDP leader and prime minister in 1974 was less a reflection of the career potency of the environmental agency post than it was a reflection of Miki's factional strength in the party and his image as being more ideologically centrist than either of the presumed frontrunners, Ohira and Fukuda.

The Environmental Agency on the one hand has to be able to make its voice heard in the policy committees of the LDP, the cabinet, and the Diet. On the other hand, it must serve as a bridge between the various empires of the

Steps," a speech by Oishi reported in *Japan Report* 18, no. 13 (July 1972): 1–3.

66. "World Environment Newsletter," *World*, 24 October 1972, p. 43.

67. A description of intraparty jockeying for power among Takeo Miki and other LDP factional leaders is included in Robert Shaplen's "Letter From Tokyo," *New Yorker*, 20 May 1974, pp. 114–15.

fragmented bureaucracy. Its creation was coupled with the elimination of environmental offices in most other departments. But paper consolidation is not the same as operational consolidation, and the agency still must overcome the prerogatives of older departments armed with large budgets and established clienteles and policy priorities.

Within the Environmental Agency itself are four bureaus: planning and coordination, nature conservation, air quality and water quality. The agency also oversees the operations of three ancillary councils dealing with environmental pollution, parks, and wildlife. This assortment of formal responsibilities is not assurance of bureaucratic effectiveness, any more than it has been for American or Soviet environmental agencies. It will take time and careful cultivation for the Environmental Agency to produce the sort of networks of interpersonal relations and public prestige that has bestowed such influence on older Japanese ministries. The Environmental Agency does formulate all policy recommendations dealing with environmental programs and pollution control, but it must do so by practicing the consensus-building or "piling up" art. As one observer describes the process by which the agency carefully acquires seals of approval from all affected departments, an environmental proposal might be held up for months because of powerful business leaders who are consulted by MITI bureaucrats."[68]

But the processes of policy making and environmental deterioration do not come to a standstill while a new agency accumulates its political resources. In January 1974, for instance, the agency on behalf of the government announced controversial auto emission standards.

68. Kenneth Stunkel, "Environmental Administration and Decision-Making in Japan" (Paper presented at the Northeastern Political Science Association meeting, 7–9 November 1974), p. 16.

Aimed at reducing the carbon monoxide and hydrocarbon emissions to one-tenth of their present levels by mid-1975, the Environmental Agency's standards met with opposition from Japan's auto makers, who produce nearly 80 percent of the country's automobile output. The companies, like those in the United States faced with EPA standards, called for delayed enforcement on the grounds that the technology was not yet available. However, according to an official government release, the auto makers "reversed themselves in the face of a tough government stance backed by strong public opinion." [69]

The release, not without a touch of smugness, went on to note that in fact the new Japanese emission standards were the same as those of the U.S. Clean Air Act which have failed to be implemented in the United States.[70] The Japanese Environmental Agency's apparent success in imposing stiff standards is due not to the Agency's own power, but rather to the decision of the LDP leadership to respond to growing public impatience which in turn was undermining the LDP's electoral hegemony. The auto makers and big business as a whole may have seen such surrender to public opinion as the necessary price paid to maintain the LDP's dwindling electoral margins. Furthermore, the Japanese auto makers own research and development investment in the auto-emissions field had jumped from $9 million in 1966 to $110 million in 1972.[71] Ironically, however, the bur under the saddles of Toyota and Datsun was not Japanese governmental restrictions but those being passed in the United States, a prime market for Japanese cars. Finally, there is evidence that when the businessmen appealed to their traditional protectors in MITI for help in delaying

<hr>

69. "Japanese Government to Enforce Auto Emission Standards from 1975," *Japan Report* 20, no. 8 (16 April 1974): 3.

70. Ibid.

71. "Japan: Limits to Growth," p. 12.

the imposition of emission standards, MITI officials advised them instead to "toe the line on the environmental issue." [72]


## CONCLUSION

Among the world's nations, Japan has been remarkable for its sensitivity to external pressures and for its capacity to make adjustments while retaining its cultural identity. Perhaps it is this traditional sensitivity, plus the postwar reliance on material prosperity as the surrogate for international power, that have made environmental problems so politically prominent in Japan. For Japan, more than almost any other nation, has awarded pollution political centrality. This is one of the few political systems in which a ruling party might lose power because of citizen discontent with environmental hazards. This is one of the few in which pollution has been a major force behind redefinition of national identity and foreign policy.

Political saliency and centrality, however, do not guarantee that power relationships will be so restructured that environmental protection will become automatic. "Japan, Inc." still exists. And it may well be that stricter antipollution measures and long-range environmental priorities will be instituted without dismantling this potent alliance between the Liberal Democratic party, big business, and top-level civil servants. Increasing willingness of Japanese citizens to take direct action; increasing ability of the Socialist and Communist parties to use antipollution platforms in gaining local offices; increasing discontent among well-educated, ambitious Dietmen within the LDP itself with "geisha house" politics; and, finally, increasing opposition to Japanese over-

---

72. Ibid., p. 13.

seas investments by nationals in Southeast Asia—all could compel the current political elite to adopt new priorities and new development formulas in time to avoid being toppled. The Japanese political system would be changed—local politics would become more vital, citizen apathy lessened, intraparty splits less personalist— though the regime remained the same. Without the environmental issue, these changes in Japanese politics might not have occurred.

On the other hand, environmental policies are still the product as well as a possible shaper of Japanese political processes. In the July 1974 upper-house election, for instance, when Tanaka was still prime minister, as director of the Environmental Agency, Takeo Miki played the role of intraparty faction leader, not that of spokesman for an issue stance. For Prime Minister Tanaka challenged Miki's party position by sponsoring an upper-house candidate in Miki's own home province of Tokushima to run against Miki's hand-picked candidate. Commentators did not read this contest as a conflict over government pollution policies, but as an indication that in the next LDP convention Miki would challenge Tanaka for the party leadership.[73] When, several months later, Miki became prime minister, his public calls for reassessment of Japan's no-holds-barred dedication to economic growth cheered environmental activists. Still, they were quite aware that Miki would be constrained by the traditional necessity of any LDP leader to balance intraparty factions, often at the sacrifice of innovative public policy.[74]

73. "Intelligence," *Far Eastern Economic Review*, 8 July 1974, p. 5.

74. Koji Nakamura, "Japan: Miki, the Puppet on Fukuda's String," *Far Eastern Economic Review*, 20 December 1974, p. 14.

# POLLUTION POLITICS IN BRITAIN

## CHAPTER 8

## BRITAIN AND JAPAN

Britain's political system has stood as a model for countries throughout the world. Britain appears to have discovered the formulas for civility and stability in politics. Until recently, they in turn ensured Britain economic and political power internationally. At the core of the British system was a political culture that respected individual rights to representation and expression. The structural counterparts were competitive two-party elections, party government, and collective responsibility in the cabinet.

Environmental politics reveal some of the discrepancies between the model portrayed in so many American

**264**

and colonial textbooks and the reality of British policy making. In practice, majority-party control of government is limited by the domination of an elitist civil service in many policy realms. Similarly, the necessities of modern campaigning and government expansion have made the prime minister more than merely a "first-among-equals" in the cabinet. Third, party organizations alone have been inadequate guarantors of popular representation, that role now taken over in many instances by powerful interest groups who enjoy special access to government policy circles and close out the general public. Finally, the traditional ties between the British citizen and public authority have been weakened by the increasing incompetence or even irrelevance of local government.

Environmental politics provides an analytical window through which to see the modified British model. But it has also been an avenue for British reformers themselves, who have used the pressures of environmental hazards to push through changes that will restore accountability, competence, and relevance to the nation's political system.

At first glance, the British and Japanese political systems seem to share much in common. In the middle of the 1970s both were plagued by uncertainty and discontent. Both will cope with these profound problems within societies more integrated than most in the world and within governmental structures that have a high degree of stability. In each, environmental and other controversies will be handled by elected parliaments and party-organized regimes. In each, a prestigious civil service plays a strategic role in defining and implementing national policy. And in Britain as well as Japan the cultural preference for consensus rather than confrontation encourages officials to seek big-business cooperation through regular consultation rather than sanctions. Finally, in each system local governments are ill equipped to cope with the emerging needs of their own constituencies.

Behind these similarities, however, are important differences. Environmental issues in Britain will not experience handling identical to that in Japan. Party competition has been more effective in Britain, allowing for regular alternatives of Conservative and Labour administrations in power. The result is that environmental neglect is not perceived as the fault of a single party as it is in Japan. This rotation of parties in office also has meant that Britain's organized labor has enjoyed more access to government policy makers than have unions in Japan. Consensus politics is not so strictly synonymous with big-business privilege as in Japan.

Internally, too, British and Japanese political parties differ. Both suffer from intraparty splits. But instead of Japan's personalist factions, Britain's Conservative and Labour parties are divided internally along ideological lines. The result is that issues, such as the extent of environmental control, are more likely to play a role in intraparty maneuverings in Britain than in Japan.

While the bureaucracies of both Japan and Britain have been charged with elitism and political intrusion, these accusations are more stinging in Britain where civil servants' legitimacy is more dependent on serving their popularly elected "masters." This requirement, plus the alternation of parties in office, has encouraged an ethic of strict partisan neutrality and a clear differentiation between politicians and civil servants. By contrast, the Japanese bureaucracy and ruling LDP are closely intertwined, with the party leadership regularly intervening in bureaucratic promotions to guarantee the ascendancy of officials sympathetic to the party and with senior bureaucrats routinely retiring in order to run for political seats. Consequently, though the closed character of policy formation frustrates environmentalists in both Britain and Japan, it poses a greater barrier in the latter.

Local governments are underdeveloped in Britain and Japan, but their revitalization carries different political significances for environmental advocates in each. Japan has never had a strong tradition of local responsibility, and it is unlikely that environmental authority will be greatly decentralized at the expense of power in Tokyo. But elections in Kyoto, Tokyo, and other municipalities have shown that local politics can heighten the pressure on the central regime to take environmental initiatives if only to prevent opposition parties from using local grievances as a springboard to otherwise unattainable national power. By contrast, British local politics has elicited less and less citizenry interest. Yet environmentalists must devote considerable attention to local government because traditionally that level of the system has had primary responsibility for ensuring public health. Thus, while in Japan local politics are salient for environmentalists because of the impact the LDP–leftist party competition has on policy, in Britain local politics are crucial because of the authority vested at that level. Thus local governments' inadequacies pose a more serious problem for British policy makers.

Lastly, the sense of national unease apparent in each of these countries stems from quite contrary sources. Japanese policy makers are faced with the problems of too much success, too much potential political power internationally, plus too much actual economic growth at home and abroad. Environmental issues gain critical importance in Japanese politics because they are such visible manifestations of these internal and external dilemmas of success. British politicians and civil servants, on the other hand, must cope with the problems of a post-Empire slide from power internationally as well as a post-industrial revolution economic obsolescence. In the

British context, environmental hazards are likely to reflect not the price of an economic boom, but the price of not converting to more sophisticated and competitive technology earlier. While in Japan today environmental activists are boosters of a necessary "slowdown" game, in Britain they are apt to be seen as spoilers in a "catch up" game.

## A PERIPHERAL ISSUE

Britain's Clean Air Act of 1956 came seven years before comparable legislation in the United States and preceded air-quality regulation in countries such as the Soviet Union and Japan by even longer. The act was inspired most immediately by the killer smog of 1952 which brought London to a standstill.[1] The legislation had the active support of a dedicated interest group, the Smoke Abatement Society (now the National Society for Clean Air), and stirred up considerable controversy in the House of Commons and the press. This act was followed in 1961 with the Rivers Pollution Act. Neither was the product of widespread public mobilization. Britain's environmental movement came later.

New environmental organizations, new journals, television specials, and specifically environmental planks in Conservative and Labour platforms had to wait until the late 1960s. The catalysts were the 1967 wreck of the supertanker *Torrey Canyon* which spilled 30,000 tons of crude oil into the English Channel and the growing momentum of the environmental movement in the United States, from which British activists borrowed so many of their organizing and publicizing techniques. In 1969 both the Conservative and Labour parties inserted environ-

1. See chapter 2.

mental pledges in their annual platforms. In 1970 Europe witnessed its first Conservation Year. In Britain the First Royal Commission on Environmental Pollution was established in the same year. But in the spring election of 1970 which put Edward Heath's Conservative party in power, the environmental issue did not play a major role in parliamentary campaigning. Nevertheless, one of the initial decisions of the Heath government was to create a new "superministry," the Department of the Environment, in October 1970, making Britain the first country to have such a critical cabinet-level department.

The first comprehensive blueprint for coping with a nation's environmental problems also came from the British. Supported by the Royal Society, and published in the respected journal *The Ecologist* in January 1972, the "Blueprint for Survival" set out principles that should guide Britain's future development:

1. Economic growth and the GNP should be abandoned as the measure of the nation's development.
2. Britain, one of the world's leading "cheap food" importers, should reduce its population to the point where it can be fed by the country's own agricultural capacity.
3. Britain should stop building roads and transfer resources to restoring railway lines and canals.
4. All usage of nonreplaceable raw materials should be punitively taxed.
5. A "polluter must pay" rule should be implemented whereby companies that cannot recycle their wastes should be taxed out of business.
6. Decentralization, rather than the current centralization, should be the plan for industrial development.
7. In foreign policy, Britain should halt its support

of the chemically dependent "Green Revolution" and stop trying to bring modern "civilization" to such remote places as the Amazon.[2]

Still, the paradox of British environmental politics remains: Britain has been ahead of most nations in recognizing and responding to environmental deterioration, but the environmental issue has never achieved the electoral or ideological centrality it has had in Japan or, to a lesser degree, the United States. There may be several reasons for the apparent peripherality of the issue in British politics. First, the country's economic lag has been too worrisome to permit the sort of concern with the hazards of growth that have dominated debate in more dynamic nations. Though the very slowness of Britain's growth could possibly keep British voters sensitive to what is to be *lost* should the economy expand too rapidly. Second, the style of British politics dampens mobilization and confrontation. For instance, the public distress over the deaths caused by the London killer smog did not translate into a durable public movement. The 1956 Clean Air Act iself discouraged lasting political mobilization with its routine formula for government regulation that placed heavy reliance on bureaucratic solutions and permitted little use of penalties or fines to curtail environmental offenders. Testimony of this noncoercive approach to pollution control is the fact that the Alkali Inspectorate, which has jurisdiction over most of the major industrial processes, prosecuted only three cases during the period 1920 to 1967.[3]

---

2. See the *Ecologist*, 13 January 1972. A summarization of the blueprint and a description of the authors, Edward Goldsmith and Robert Alein, appears in Lewis Chester, "A Manifesto for Man," *Sunday Times*, 16 January 1972.

3. Howard A. Scarrow, "The Impact of British Domestic Air Pollution Legislation." *British Journal of Political Science* 2, pt. 3 (July 1972): 282.

The lack of steady mobilization and the erratic character of political attention to environmental problems is reflected in British media coverage. Figure 8.1 shows the ups and downs of air pollution stories in at least one British paper, the *Times* of London, between 1946 and 1970.

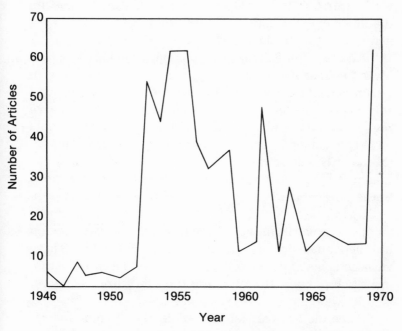

*Figure 8.1.* **Number of Articles on Air Pollution in the *Times***

*Source:* From L. T. Foster, "Some Measures of Public Interest" (Paper presented at the Man and Environment Commission Symposium of the International Geographical Union, Calgary, Canada, July 1972), fig. 2.

Another indicator of public interest in one environmental issue, air pollution, is the support given to the National Smoke Abatement Society, one of Britain's oldest and most politically visible environmental lobbyists. There has been an overall increase both in the money contributed to the Society and the number of contributors since 1929. Once again, however, the pattern has not been steady. Peaks occurred during the political controversy that followed the London fog and more recently during the European Conservation Year. On the other hand, the data for this interest-group's support are far less erratic than that for press coverage, suggesting the newspaper attention to pollution issues is far from a determining factor in British public opinion.[4]

All too often British society is portrayed as remarkably homogeneous. In actuality, it is a diverse society in which accents can be pinpointed county by county and social class by social class, in which tastes and styles vary regionally, and in which racism and even separatism now have spokesmen.[5] Such diversity means that environmental hazards will be treated more seriously in some areas than in others. Likewise, there have been mixed reactions to the government's air-quality-control measures. A principal variable affecting popular and local government compliance has been the degree to which the given region has depended on coal mining for its livelihood. These variations also underscore the degree of decentralization in British politics; Britain is not the centralized system it is often thought to be by foreigners.

Since the 1956 act left the implementation of smoke-control regulation largely in the hands of local governments, this diversity of reactions to air pollution is reflected in the unevenness of implementation of the national act's provisions. If regional cultural differences played no part in implementation, then those communities with the objectively highest ratings of air pollution

---

4. Foster, "Some Measures of Public Interest."
5. For a provocative study of the patterns of taste differences in Britain, see D. Elliston Allen, *British Tastes* (London: Panther Books, 1969). Analyses of ethnic and regional cleavages in Britain are included in Enloe, *Ethnic Conflict and Political Development* (Boston: Little, Brown, 1973).

would be the first to adopt the central government's air-quality regulations. This rational pattern does not fit the British reality. Those areas with the blackest skies and highest incidences of respiratory disease were those whose residents for generations had depended on mining companies for jobs and for traditional subsidized distributions of coal. In these mining regions, the coal fire has had emotional as well as economic meaning. These same communities were among the most limited in financial resources, another factor slowing the adoption of control measures.[6] More affluent British communities are likely to recognize the air pollution issue and act upon it not only because they are less likely to be dependent on "coal culture" but also because their local councils and health departments have the resources available to tackle the problem.

For example, in the coal-mining region of Edinburgh, Scotland, citizens distrusted or disliked smokeless fuel. Referring to cleaner fuels such as natural gas, they expressed a fear of gas explosions, of the increased price of smokeless fuels, and of a loss of homey open fires. In the late 1960s over 40 percent of Edinburgh respondents denied the presence of air pollution in their own neighborhoods at any time. This despite the scientific measurements which recorded substantial pollution levels over the entire city at least on some days of every year.[7] Not surprisingly, pollution issues rate well below other political concerns in this traditional mining area. When asked

---

6. L. T. Foster, "The Adoption of Smoke Control Areas" (Paper presented at the Man and Environment Commission symposium of the International Geographical Union, Calgary, Canada, July 1972). See also Foster's "The Adoption of Clean Air Legislation: The Response of Municipalities to the Clean Air Act of 1956" (Ph.D. dissertation, University of Toronto, 1975).

7. Douglas Billingsley, "Public Response to Air Pollution and Control Measures in Edinburgh" (Paper presented at the Man and Environment Commission symposium of the International Geographical Union, Calgary, Canada, July 1972), p. 4.

to rate local issues that they would consider "serious" or "very serious," the Edinburgh respondents in 1971—the year *after* the Royal Commission, the Europe Conservation Year, and the creation of the environmental super-ministry—placed air pollution far down the list, below vandalism and alcoholism.[8]

Low saliency of the pollution issue continued to hamper air pollution control. By 1969 only half of the 324 designated "black areas" (areas of objectively high pollution levels) in England had been fully covered by smoke-control orders. Many areas are not projected to be covered until as late as 1987 or 1989, thirty years after the Clean Air Act.[9] Local authorities fear that adoption of central government regulations will drive away industries at a time of economic hardship. To make the public more actively concerned about pollution and willing to pay the price of effective control, the direct costs to the nation of treating bronchitis and corrosion, painting and decorating, as well as depreciation of buildings and damage to textiles could be publicized. Such a strategy was proposed by the Beaver Commission which investigated the 1952 smog. It estimated that in terms of "loss of efficiency" air pollution cost the British economy some £250 million annually. By the late 1960s the Ministry of Technology guessed that the figure was closer to £350 million per year.[10]

### BRITISH ISSUE NETS

The saliency of environmental issues in Britain reflects not their centrality but their *strategically periph-*

8. Ibid., fig. 10.

9. John Barr, "Air Pollution," *New Society*, no. 369 (23 October 1969): 647.

10. Ibid.

*eral* position. They have been significant dimensions of four prime issues facing Britain during the mid-1960s: (1) restructuring of local government for the sake of revitalizing local citizen participation; (2) institution of town and county planning in order to cope with spreading urbanization; (3) improvements in housing, an issue highlighted by the racial and ethnic conflicts in Northern Ireland and England; (4) the modernization of domestic and international transportation facilities to cope with the increasing mobility of Britons and the need to compete in international markets (see fig. 8.2).

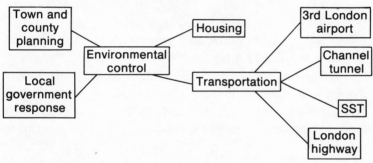

*Figure 8.2*

Issues attracting the most mobilized attention of British environmentalist organizations have been those relating to transportation policy: the London highway; the third airport for London proposed for a site at Maplin; the rail and road tunnel proposed to run under the English Channel between Britain and France (referred to as the "chunnel"); and finally, the Concorde, the supersonic jet that has been developed at an enormous cost by France and Britain.

In each policy question the central government has primary jurisdiction. In each, the interest of government and industry are intertwined, hastening the politicization of the decision. In each, the government must weigh costs to the economy: inflationary in terms of government expenditure or depressing in terms of loss of jobs for British workers. In all but the London roadway controversy, foreign policy repercussions are crucial, especially the impact on Britain's status in the European Common Market should the Concorde or chunnel be abandoned.[11] The publicity and organized lobbying that these issues aroused resulted in mixed results for environmentalists. The Conservative government approved the six-lane road through residential sections of London in February 1973, but opponents allied behind the anti-road Labour party candidates, giving them a victory in the Greater London Council elections of 1973.[12] The Heath regime also persisted with the Concorde SST project, fearing the loss of 55,000 jobs should it be terminated. But the new Labour regime under Harold Wilson at least temporarily halted work on the Concorde after taking office in 1974, though salesmen for the British Aircraft Corporation and its French counterpart continued to shop for clients around the world.[13] The effective anti-Maplin lobbyists managed

---

11. For a detailed account of the environmental implication of the chunnel, see "Bore of the Century?" *New Scientist*, 11 October 1973, pp. 92–113. On the "Third London Airport Issue," see Maplin, "How the Mind Was Made to Stick," *Guardian* (Manchester), 23 June 1973, p. 10; John C. U. Adams, "London's Third Airport," *Journal of the Royal Geographic Society*, November 1971, pp. 465–504; "Aground on Maplin Sands," *Economist*, 30 June 1973, p. 87. Regarding the six-lane ring road planned for London, see Peter Willmott and Michael Young, "How Urgent Are London's Motorways?" *New Society*, 10 December 1970, pp. 1036–37.

12. This electoral analysis was provided by Professor Janes Dunn, who is engaged in research in Britain's transportation policies, September 14, 1974. Regarding road politics in London see also Simon Jenkins, "The Politics of London Motorways," *Political Quarterly* 44, no. 3 (July-September 1973) and *New York Times*, 20 February 1973.

13. There is already a controversy brewing in the United States over whether to allow the Concorde to land regularly at American airports. In 1974 the first Concorde SST did land at Boston's Logan Airport despite local environmentalist objections. Regarding the Concorde debate in Britain, see Andrew Wilson, *The Concorde Fiasco* (Middlesex, England: Penguin Books, 1973).

to persuade the government to shelve its third-airport plans in 1973, but opponents of the Channel tunnel faced a persistent government determination to proceed with both an auto road and a rail link, whereas environmentalists argued that only the rail link is needed.[14] In January 1975, Harold Wilson's Labour government did decide to halt work on the channel tunnel, but based its decision on Britain's economic stringencies, not on any environmental doubts.

## DEVELOPMENT AND ENVIRONMENT

One factor that has breathed new life into ethnic nationalism in Britain has been the economic disparity between England on the one hand and Scotland and Wales on the other. As part of the local government reforms, there has been growing support for permitting these regions greater local autonomy. However, the Arab oil boycott and its reduction of energy resources for Britain in a period of severe economic strain made central government officials all the more aware of the value of those two regions: Wales for its coal and Scotland for its coal and offshore oil and gas. But while officials in London were conceiving of ways to exploit these valuable energy resources for the sake of the entire country, local nationalists saw the resources as their best means yet for bargaining for autonomy and for increasing their depressed areas' portion of the British economic pie.

---

14. Mark Arnold-Forster, "Maplin Shelved, Chunnel Go-Ahead," *Guardian* (Manchester), 23 June 1973. The next year the Labour government's secretary of state for trade and industry announced to the House of Commons that the Maplin project would be halted due to declining air traffic. Environmental factors were not mentioned. *New York Times*, 19 July 1974. On the Channel tunnel controversy over whether to have both as road and a rail link the British Road Federation, one of the strongest lobbies in DOE affairs, was able to prevent a government switch to merely a rail tunnel.

The dispute over control of Scotland's offshore North Sea oil had much in common with the American dispute over the Alaskan oil pipeline. Both involved territorial units that had special identities, relative remoteness from central governments, and ethnic privileges. Both debates were international in scope, the Alaskan pipeline involving U.S.–Canadian relations, the North Sea oil drilling involving other countries, such as Denmark, with offshore claims. Both raised questions about the hazards deriving from fragile environmental conditions, the Alaskan tundra, or the stormy sea. Lastly, both involved multinational corporations such as Exxon, Shell, and British Petroleum.

In the Alaskan case, national interest groups such as the Sierra Club utilized the environmental impact statement provisions of federal law to do battle within the central bureaucracy and in the courts, in addition to mobilizing sufficient public opinion outside Alaska to make a respectable showing in a congressional vote. By contrast, in Scotland the environmentalists never seemed to catch the imagination of the larger British public or even the Scottish majority. They mounted public campaigns, aimed largely at the local councils in coastal towns most immediately affected by the oil boom. They exerted pressure on the Department of Trade and Industry, the secretary of state for energy, and the secretary of state for Scotland, the London-appointed minister with principal authority over Scottish affairs. Only the Liberal party warned against unbridled exploitation of the oil. No major faction within the SNP, Conservative, or Labour party took up their banner. Not until 1974, when the ecological implication of the oil boom became clearer to Scotsmen themselves and when the newly elected Labour government began to take SNP rivalry at the polls seriously, did environmental issues gain prominence. In August, 1974, the Labour government turned down a proposal to build North Sea oil platforms at the scenic coastal village of Drumbuie. Furthermore, Labour's secretary of state for energy announced plans to nationalize all other oil sites on the Scottish coast in order to protect the environment. Scottish nationalists, however, suspected that London was even more eager to protect profits for itself.

Both Labour and Conservative parties made oil profits an issue in the February and October 1974 elections. Labour called for nationalization and criticized the Heath government for giving away too much to the multinational corporations. But the party that reaped most from the North Sea oil issue was neither the Conservatives nor Labour but the nascent Scottish Nationalist party. Oil became the best issue that the SNP had yet had to make their case for greater autonomy for Scotland. With its call for Scottish control of the North Sea resources, the SNP won 7 seats in the 1974 parliamentary elections, enough to give them swing power to make the new minority government of Wilson's Labour party shelve temporarily their talk of nationalization. The election did little if anything to raise the saliency of the environmental issue. But in the year's second election, in October, the SNP did even better, increasing its delegation in Commons to 11 seats. And this time environmental concerns began to attract more attention. Still, Scottish nationalists look to profits, not unspoiled nature, to give them a springboard for regional autonomy.[15]

The North Sea oil dispute is marked by some of the conditions that seem to keep environmental concerns off center-stage in Britain. First, it was not a matter which

15. Analysis of the North Sea oil politics and Scottish nationalism can be found in *New York Times*, 13 August 1974; Alistair Reed, "Letter from Scotland," *New Yorker*, 7 October 1974, pp. 76–116. *Washington Post*, 2 September 1973, 21 October 1973, and 2 December 1973; "Where to Build Oil Platforms," *Guardian* (Manchester), 5 June 1974, p. 16; Neal Ascherson, "The Highland Oil Fling," *Observer* (London), 3 March 1974.

shook some basic national myth or "dream" such as Soviet man-as-producer or the unlimited frontier or national redemption through material prosperity. Second, Britain's problems of national economic failure are too acute to permit more than a fragmentary concern with environmental issues. Finally, sufficient steps have been taken by politicians and career civil servants early enough to avoid repeat of those environmental catastrophes which galvanize nationwide movements.

## GROUPS AND PROCESSES
## IN THE "COLLECTIVIST AGE"

Environmental politics has arisen in what Samuel Beer terms Britain's "Collectivist Age." [16] Beginning at the turn of the century and coming into full bloom after World War II, this era of British political development is marked by a growing involvement of government in the management of the economy and a blurring of lines between government and producer groups, whether they be manufacturers or trade unions. Since the 1940s both business and labor groups have gained a direct access into the circles of decision making in London, an access deemed legitimate and efficient. This access for producer groups is accepted by both Labour and Conservative parties, despite their contrasting bases of electoral support.

As the 1956 Air Control Act suggests, British policy makers prefer to achieve their ends through consultation, gradualism, and anticipating opposition through prior policy modification. But this consultative pattern and allowance for group access does not result in a system as fragmented as the much larger American system. Belief in controlling excesses of power through deliberate frag-

16. Samuel H. Beer, *British Politics in the Collectivist Age* (New York: Vintage Books, 1969).

mentation of power, plus a belief that the best policy is the product of open conflict, has given environmental politics its special flavor in the United States. In Britain, by contrast, power is more concentrated in the capital, while party discipline, cabinet authority, and civil service continuity make the policy process less disjointed. However, power in Britain "is not the exclusive property of any one of its components, even the most influential." [17] For instance, the prime minister is indeed powerful, but only when he has the backing of his own party and the acceptance of the country's chief interest groups. If, as in early 1974 during the disastrous coal miners' strike, a prime minister loses the confidence of these groups, he is likely to resign and call for new elections. If, as in the case of Harold Wilson, the prime minister and his party are elected by less than a majority, he is also likely to call for early elections to strengthen his mandate. For the British political system is designed to gain widespread compliance without alienating major groups and without depending on coercive sanctions.

The contrasts between British and American policy making were underscored for environmentalists when American reformer Ralph Nader visited Britain in 1971. Nader's message was that by relying on consultation and cooperation the ordinary British consumer was tolerating closed relationships that narrowed the definition of the public interest. A British group created after Nader's tour, the Public Interest Research Centre, not only took on Nader as an overseas adviser, but also adopted Nader's tenets for countering this sort of closed and "cozy" policy system. The Centre pledged that it would work for more public involvement in decisions which determined the quality of their lives, such as consumer and environmental decisions. It argued that this goal would be impossible

17. Karl W. Deutsch, *Politics and Government* (2nd ed.; Boston: Houghton Mifflin, 1974), p. 417.

unless a more open policy-making and administrative process was adopted, not only in government but in corporate business as well.[18]

One of British journalism's most environmentally conscious practicioners, Peter Hall, visited the United States just prior to Nader's British tour and noted that whereas Americans assume that there is no "Rousseau-like general consensus" in society and that resolution of naturally differing individual interests will have to come out of battles in the voting booth and courtroom, the British put more faith in the collective establishment. This is an establishment at the center, leaving little room for grass-roots populism of the sort that Nader has depended upon.[19] Hall predicted that exporting Naderism to Britain would prove difficult, but perhaps necessary, if the environmental movement is to have any genuine impact on national policy. Hall pointed to the Nuclear Disarmament movements of the early 1960s and the more recent British civil rights movement stemming from the disputes over discrimination against Caribbean and Asian immigrants as fertile soil in which Nader-like groups might develop. "Naderism, save in a few pale manifestations, has yet to arrive; it awaits a catalytic figure," according to Hall.[20] But the prospect is that in Britain too, the next decade will be a time when political innovation comes from the periphery, not the centre.

Changes in institutional roles will be required if the Nader political formula is to be exported to Britain in toto. In the United States, Nader's public-interest groups have been excluded from state and federal policy circles but have had recourse to the courts. British courts have not played this policy role.

---

18. *Times* (London), 23 October 1971. See also "Newsmakers," *Newsweek*, 1 November 1971, p. 47.

19. Peter Hall, "Going American," *New Society*, 15 July 1971, p. 117.

20. Ibid.

Acknowledging this, an official of the Nature Conservancy, a governmental environmental council, wrote to the *Observer* following Nader's departure supporting his objectives but wondering whether his methods were applicable to Britain. The writer questioned especially the Nader reliance on law and the courts as political weapons. "Sue the bastards" is a standard American conservationist technique and is particularly effective when the suit is aimed at the American government itself. But in Britain, the Conservancy spokesman notes, this sort of tactic could lead to a "dangerous polarization" between policy participants. Polarization would mean that industry or a stubborn government ministry would have to be seen as enemies, contrary to the British style of interaction. "Egged on in the witness box by a committed lawyer, otherwise responsible scientists can be persuaded to lie—or at least to exaggerate or select their evidence. It is the environment which suffers."

Confrontation and irrationality are anathemas to British political actors. Better, set up a new quasi-government board or create a Royal Commission than to trust national policy to the heat of the courtroom. The spokesman notes that in Britain, "we have adopted a different method. We have tried to cooperate with industry and to introduce effective, if often voluntary, controls." In the case of pesticide control, for instance, he finds that such cooperative voluntarism obviated danger, even if not by the most efficient method, to the extent that today in Britain alone endangered species of birds are showing some recovery. In conclusion, the writer contends that "we may not be so noisy, we may seem inefficient, but we may be more effective than the Americans." [21]

But at the same time as hints of a Nader-ist group activism appear, there are efforts to replace British

---

21. Keith Mallanby, "Letter to the Editor," *Observer* (London), 7 November 1971.

pluralist "muddling through" with more coherent, long-range planning. The consolidation of bureaucratic departments and the reform of local governments both have been directed toward making the system more amenable to planning requisites and less dependent on ad hoc sorts of decision making.

In a 1970 report on the third London airport, this call for more rational planning processes was central. The Roskill Commission concluded that the British policy system required more rational decision making, utilizing more objective techniques such as cost analysis. Supporters of the Roskill Commission warned that *organized group* activity does not necessarily promote beneficial results.[22] Furthermore, electorally responsible parties, according to this critique of current British practice, do not guarantee effective policy making. For example, although environmental issues gained entry into parliamentary electoral campaigns in 1970 and 1974, the environmental pledges of both Conservatives and Labour remained too vague for voters to make rational choices based on party stands.[23]

Citizens as well as experts have expressed growing dissatisfaction with the present mode of policy formation. A 1973 government-sponsored study concluded that "of all the aspects of life which were surveyed, the state of democracy in Britain registered the lowest level of satisfaction. Most people felt that although there is plenty of freedom of speech, as voters they have very little influence on the way the country is run." [24] The greatest dissatisfaction was registered by the young and the better educated.[25] The second annual report of the Royal Com-

---

22. Lincoln Allison, "Politics, Welfare and Conservation, A Survey of Meta-Planning," *British Journal of Political Science* 2, pt. 4 (October 1971): 445.
23. Ibid., p. 446.
24. "Social Trends: Statistics of Satisfaction," *Economist*, 15 December 1973, p. 28.
25. Ibid.

mission on Environmental Pollution also concluded that there was a need for more public openness in environmental policy processes.[26]

Three conditions are most often cited by environmental groups as stifling public participation: (1) the paternalism of the civil service-dominated departments who hold the purse strings in London, (2) weaknesses of local governments and public apathy about local affairs due to the failure of local authority lines to match the services that citizens care most about, and (3) a vague though measurable discontent with the two major parties.

Policy access only for well-established groups seen as serious and legitimate in the eyes of officials excludes many environmental organizations. Reliance on voluntary compliance engenders benign tolerance toward industrial violators of antipollution laws, a tolerance that is very difficult for outside groups to challenge effectively. One commentator has called this Britain's "Nanny knows best" approach to pollution. Jon Tinker, writing in the *New Scientist,* contends that "some of Britain's pollution rules are better suited to an Edwardian girl's school than to an advanced industrial society." Offenders, he reports, "are taken quietly on one side by the prefects and ticked off for letting the side down. There is no need for prosecutions: the shame of being found out is reckoned to be punishment enough. Carefully shielded from vulgar eyes, pollution control operates behind a deliberate smokescreen of evasion and reticence." [27]

The Alkali Inspectorate, the local river authorities, and the Factory Inspectorate, whose responsibilities include enforcement and administrative adjustments (read, policy making) of air- and water-quality laws all enforce antipollution laws according to the same principles.

26. Royal Commission on Environmental Pollution, *Second Annual Report* (London: Her Majesty's Printing Office, Cmnd 4894, 1972).

27. Jon Tinker, "Britain's Environment—Nanny Knows Best," *New Scientist* 53, no. 786 (9 March 1972): 530.

Principle 1: Close cooperation between regulating authority and industry.

Principle 2: Each case is considered on its merits so that emission limits are fixed on the basis of local conditions rather than forced to conform to some nationwide norm.

The public is unable to evaluate the worth of these policy decisions because of the standards of secrecy which cloak so much of the British policy system. "Like an elderly nursemaid, the Alkali Inspectorate suggests that publication of *emission* figures could only serve to alarm the public, in spite of the fact that official reports on *environmental* levels (mercury in shellfish and DDT in river water, for example) have been released without mass hysteria." [28] The nanny approach discourages citizen involvement: "A nasty taste in your jam, dear? Nonsense. Nanny's looked at it very carefully and there's nothing at all wrong. Eat it up like a good boy. Nanny knows best." [29]

The legal restrictions on release of pollution information have curtailed environmental groups' involvement in both national and local decisions. A study of the decision to close down the world's largest blast furnace, at Avonmouth, revealed the short-circuits in information flows. The dispute was over whether the smelter, which would fulfill two-thirds of Britain's entire zinc requirement, was endangering the health of local residents with its cadmium pollutants and thus should be shut down. Involved in the decision were scientists for the smelting firm, local university scientists, the local planning authority, the public health offices, local councillors, the National Alkali Inspectorate, the mass media, voluntary action groups,

28. Ibid.
29. Ibid.

and ordinary citizens of Avonmouth. The study found that very few local residents infiltrated information either by contacting the press or local councillors with their complaints. Local action groups proved better informed and more willing to pass along information regarding the pollution hazards. Government scientists passed information along to others sharing their particular professions, rather than to nonscientists in their own departments or to the lay citizenry. Though the smelter was temporarily shut down, the smelter dispute in fact resulted in increased solidarity within the scientific community and thus even greater concentration of information out of the hands of organized or unorganized local citizens.[30]

A unitary system can coexist with considerable local political autonomy. This has been the case in Britain. In fact, local government reorganization has attracted so much political attention in the last decade because the British political culture places considerable positive value on local autonomy. Much of the authority for carrying out air and water pollution standards and land zoning is in the hands of local officials. Compared to American local politics, the British are more formal. Greater importance is assigned to appointed officials. There is less local interest-group activity and much more stress on nationally defined party ties. The key to the process of local

---

30. Anne Kirkby, "An Information Model Approach to Human Responses to Man-Made Environmental Hazards" (Paper presented at the Seminar on Human Response to Man-Made Environmental Hazards, Butler University, Indianapolis, August 1973). The closing of the giant smelter was only temporary. Under its new owner, Rio Tinto Zinc of Australia, the smelter was reopened. No new legislation was passed, though the smelter now exceeds standard air pollution limits set in British law. The Alkali Inspectorate and RTZ were left to work out the pollution problem on their own. More recently RTZ's environmental adviser told an environmental group representatative that the smelter no longer exceeds air pollution limits. The Australian firm has introduced new antipollution devices, in part in response to the growing environmental sensitivity back in Australia. Personal correspondence with David Withrington, International Youth Federation, 4 September 1974.

environmental decision making is the relationship between the chief officer, an appointed official, and the chairman of the particular councillor committee dealing with a given issue. In the relationship between the central government's ministry and the local authority, the chief officer is the crucial point of contact.[31]

The principal defects in British local government have been:

1. artificial and irrelevant division of town and country, a fragmentation between "urban" and "rural" that frustrates planning and implementation

2. a lack of correspondence between the "welter of authorities and the realities of peoples lives" which led to popular ignorance and indifference in local affairs

3. the increasing take-over by London of tasks that belonged to local units as the result of local authorities' inadequate financial and technical resources[32]

Local authorities have been cautious and even timid in acting against polluting industries out of reluctance to scare away firms employing local labor. They also have been dependent upon the pollution data supplied by the polluters themselves and lack the technical personnel who could challenge that data or draft meaningful regulatory revisions to cover current loopholes. Moreover, British local governments possess inadequate ratable funds of their own with which to launch innovative programs. Whereas in 1914, of the total local expenditure 25 percent came from central government grants, by

---

31. Foster, "Clean Air Legislation," chap. 5.
32. H. Victor Wiseman, ed., *Local Government in England, 1958–1969* (London: Routledge & Kegan Paul, 1970), p. 98.

1970–71 the dependency was projected to reach 57 percent.[33]

The Department of the Environment chose local authority reorganization as its key to environmental reform. Parliament passed the Local Government Act of 1972, creating a two-tier system of 38 unitary authorities and 300 local districts in England and Wales (Scotland and Ulster have more autonomy and are dealt with separately). Some observers have said that the new authorities are "too many and too small." For instance, land-use planning, previously in the hands of 150 local authorities, now is spread among 380.[34] With no regional ties, initiative, environmental policy making is likely to flow all the more into the hands of the central government's civil servants. The main beneficiaries of the reorganization appear to be the regional arms of the two superministries, the Department of Trade and Industry and the Department of the Environment, which have gained authority in the areas of water, sewage and health. The losers seem to be the metropolitan county units.[35]

## GROUP ACCESS

At the national level, environmental decision makers include:

—the cabinet and its secretariat
—top elected and civil service officials within the Department of the Environment, the Department of Industry (formerly the Department of Trade and Industry), and the new Department of Energy
—Treasury officials

33. Ibid., p. 168.
34. G. W. Jones, "The Local Government Act, 1972 and the Radcliffe-Maud Commission," *Political Quarterly* 44, no. 2 (April–June 1972): 159.
35. Ibid., pp. 160–64.

—Central Electricity Generating Board
—British Petroleum
—the National Coal Board
—British Waterways Board
—the British Steel Corporation
—British Airways
—British Railways Board
—business associations such as the Chemical Industries Association, the Confederation of British Industries, and the British Road Federation
—Trade Union Congress
—associations of local government officials

The mixed public-private boards and corporations are not unlike those vehicles through which the Japanese formalize cooperation between government and producers. "Close and ubiquitous" contacts are made either through direct consultation with the ministries or through various boards and ad hoc panels established to investigate a particular problem. Nationalized industries such as coal, electricity, and steel, the mixed boards are oriented toward economic growth. Their formalized access and policy-making authority make them formidable adversaries for any outside interest group. American groups that have sought to curtail the Tennessee Valley's pro-strip-mining activities would appreciate the obstacles facing a British environmentalist group. Acknowledging growing ability of executives in the nationalized industries to outweigh their ministerial counterparts in issue debates, it has been suggested that future governments might have to recruit less able men who would not intimidate the junior ministers nominally in charge of them.[36]

The business lobby that has had the most influence

---

36. "Keep Them Frozen," *Economist*, 5 May 1973, pp. 73–74; on the British Steel Corporation, see *New York Times*, 19 May 1974.

not only in environmental affairs but also in economic policies generally has been the Confederation of British Industries. The CBI has been the principal industrial group with which both Labour and Conservative regimes consulted. Its members include private firms and government-owned companies as well. The CBI has regional councils which serve as links between industry and the regional Economic Planning Councils and act as consultants for local governments.[37]

The CBI, however, wields power subtly. Its very moderation may be the reason it is relied upon by the government. It is rarely an initiating force behind a decision or a piece of legislation. Rather, its influence takes the form of restraining the government on the verge of taking another step in economic intervention. Between the two houses of parliament, the CBI has found the House of Lords to be the most useful arena in which to press for detailed amendments because members of the Lords have more individual independence, whereas the members of the Commons are more subject to party discipline. The three factors that may temper the success of the CBI to achieve any given policy goal are the determination of the government to pursue the particular policy, the interest orientation, and activism of other interested groups and the degree of public awareness of the issue.

On the pollution issue the CBI, for all its access to high government circles, has failed to reverse the trend toward higher costs produced by stricter control over the discharge of industrial effluents. Still, the organization has obtained certain concessions. It is in pursuing detailed changes in broad legislation that the CBI is most

---

37. W. P. Grant and D. March, "The Confederation of British Industries," *Political Studies* 19, no. 4 (December 1971): 403–15. See also F. G. Castles, "Business and Government: A Typology of Pressure Group Activity," *Political Studies* 17, no. 2 (June 1969): 160–74.

skilled, rather than in reversing a general policy trend supported by the civil service and political establishment. For instance, the CBI lobbyists have been able to affect the *timing* of the effluent laws' implementation so as to give the industries more time to adjust before meeting stiffer standards. CBI lobbyists also won modifications in the implementation of the law. The CBI can offer ministries valuable technical data and staff work. In return for such cooperation the Ministry of Housing and Local Government in 1966 issued an official circular advising river authorities to take account of the costs to industry of effluent control. The CBI has access to the new Department of the Environment as well as official standing in joint committees with the Water Resources Board, the Association of River Authorities, and the Local Authorities Association.[38]

Interest groups that do not enjoy official status do not benefit from the government consultative system that is the hallmark of Britain's "collectivist age." An official of the British branch of an international environmental group says that in Britain, as well as other larger European countries in which his organization works, "environmentalists are considered to be troublemakers, rocking the boat." [39] Lack of credibility in the eyes of politicians is exacerbated when, as in Britain, the environmentalist lobby is split among a number of different groups. There is no equivalent of the CBI or the Trade Union Congress which can claim legitimate access in the policy process on grounds of representing an entire interest sector. Another officer in the same organization confirms this observation by contrasting their operations in Britain with counterparts in the United States: "For our

38. Grant and Marsh, "Confederation of British Industries."
39. Correspondence with the author, from David Withrington of the International Youth Federation for Environmental Studies and Conservation, 23 July 1973.

part, in Britain, the idea of the U.S. pressure group manipulating Congressional hearings and wielding a power quite unknown in this country is something one learns in the first term [of a British university] studying the U.S. Constitution." [40]

Some of the obstacles that British environmental lobbyists confront in making their influence felt in national policy formation have already been mentioned. There is: (1) the secrecy that characterizes much decision making; (2) the reliance on several well-endowed groups as the chief consultants on issues; (3) the reliance on civil servants in policy making, which makes public exposure of conflicts less likely. In addition, environmentalists are hampered by British journalism, which is less oriented toward muckraking revelations and permitted less access into administrative departments. It is far easier for interest groups that lack official status to employ their limited resources in the policy process if they can gain information from the press about who is backing what policy and what are the timetables for crucial decisions.

## PARLIAMENT AND PARTIES

There has been much talk in Britain about the demise of Parliament under the pressure of government to deal with technical and complex policy questions. Some of

40. Correspondence with the author, Julian Cummins, secretary-general of the International Youth Federation for Environmental Studies and Conservation, 18 October 1973. By contrast, British environmental groups, working in alliances such as "Transport 2000," have indeed been influential in shaping government restrictions on the weights and routes of pollution-producing heavy trucks. See Richard Kimber et al., "British Government and the Transport Reform Movement," *Political Quarterly* 25, no. 2 (April–June 1974): 190–205. Case studies of environmental campaigning are collected in Richard Kimber and J. J. Richardson, eds., *Campaigning for the Environment* (London: Routledge & Kegan Paul, 1974).

Parliament's characteristics especially limit environmental lobbying. Environmentalists as well as business lobbies are hampered by the traditional party discipline guiding the actions of British M.P.s. The Labour and Conservative parliamentary delegations are not monolithic, and any prime minister who ignores the feelings of his backbenchers risks losing parliamentary control. But M.P.s are more beholden to their respective party central headquarters and parliamentary Whips than are the fragmented American congressmen or the factionalized Japanese Dietmen. A British M.P. lacks the staff or even the office space to develop independent policy stances. The Parliament also does not have the vigorous policy-oriented committees that become the target for so much lobbying pressure in the United States. Finally, as in Japan, legislators have relatively little opportunity to submit their own private bills; instead, they are reactors to bills drafted and submitted by the majority party's leadership.[41] All of these shortcomings have been made more blatantly visible as M.P.'s have had to face the new "superdepartments," such as the Department of the Environment, which command even greater technical resources and place even more complex policy choices before Parliament.

For instance, the British Friends of the Earth in 1973 drafted an Endangered Species Bill. They had the assistance of a parliamentary draftsman and hoped to get the bill introduced as a Private Member's Bill. Then FOE discovered that the Conservative government was committed to drafting its own bill and so withdrew their legislation, thinking it more efficient to devote its energies

---

41. Dick Leonard, "Private Members' Bills in Decline," *New Society*, 25 November 1971, pp. 9–10. Leonard, a Labour M.P., contends that private bills are even more curtailed under the new Conservative government than under the Labour.

to seeing that the government's legislation was effective. This role meant the FOE lobbyists would keep watch over the Department of Education and Science (DES) and the interdepartmental working committee set up to study the implications of the new international convention protecting endangered species. FOE representatives worried about the ministerial timetable, hearing that the DES expected it would take at least a year before legislation would be actually sent to the House of Commons. To prod the administrators and cabinet into action, the FOE urged its own members to write their local M.P.s asking them to call for legislation at an earlier date.[42] Eight months later, the FOE reported to its members that it had gone ahead in the face of ministerial foot-dragging to sponsor its own bill. The Private Member Bill on endangered species was introduced by a member of the House of Lords, since members of the upper house are more independent. But months of legislative and bureaucratic lobbying and monitoring would still be needed to ensure that the bill finally became law.[43]

Both major parties have adopted environmentalist plans, though each assigns priority to other policy goals, especially anti-inflation and economic growth. The result is that environmental regulatory bills have been introduced regularly under both party regimes and administrative reorganizations to facilitate environmental planning have been promoted by both parties when they occupied Number 10 Downing Street.

However, the two major parties have suffered a loss of public confidence in recent years, critical in a system in which party control is the linchpin of policy formation.

---

42. Angela King, "Friends of the Earth: Endangered Species Campaign," *Ecologist* 3, no. 6 (June 1973): 233.

43. Peter Wilkinson, "Friends of the Earth," *Ecologist* 4, no. 1 (January 1974): 34.

The weakness of Conservative and Labour support has been reflected in electoral victories of minor parties, especially the Liberal party and the Scottish and Welsh nationalist parties. The Liberal party has led all other parties in adopting programmatic priorities that dovetail with environmentalist concerns. The Liberals campaigned in the early 1970s on a platform of revitalized local government and a new definition of national interest that did not equate the quality of life with economic growth. Some Liberal delegates at the party's annual conference also wanted a change of tactics. A young Liberal from Manchester told his fellow delegates that the country needed a campaigner like Ralph Nader who was not afraid to use publicity and direct action to expose the shortcomings of the auto companies. He warned that since pollution was now the "in" thing to oppose in Britain, some companies were taking public stances against pollution while continuing to be among the worst polluters.[44]

The Liberals have done well in local council elections, but nationally they won more votes than seats and so remained an influential factor under Wilson's minority Labour government in 1974 without possessing enough power to take the initiative in drafting the legislation. The shortcomings of the Liberal party derive not only from the unrepresentativeness of British electoral winner-take-all rules but from the relative narrowness of their popular backing. Like the environmental interest groups, the Liberal party has mainly middle-class backers who fall between the two ideological stools of working-class democratic socialism and Tory paternalistic conservatism. Yet the Liberals have put up a good enough showing in elections to have made both major parties more prone to adopt environmental protection policies.

---

44. *Daily Telegraph* (London), 16 September 1971.

## POLICY MAKING IN WHITEHALL

Paralleling Parliament's eclipse is the concentration of policy-making power in the ministries and cabinet. Britons wonder whether the prime ministership is taking on the character of the American Presidency. One should not underestimate the continuing importance of collective responsibility within the cabinet, the strength of party organizational ties, and the influence of top-level career bureaucrats who hold positions that in the United States would go to political appointees of the President. Yet the prime minister today is more than the traditional "first among equals." National campaign styles in an era of mass media have placed heavier emphasis on the office of prime minister, although the voters continue to cast ballots only for their own local M.P. The personal staff of the prime minister, separate from the cabinet staff, also has expanded. Though British leaders are anxious to point out that it falls far short of the mammoth staffs now attached to the White House.[45] The cabinet too has been strengthened. It has a secretariat, which prepares cabinet agenda and gives departments their "marching orders" in the form of policy decisions sent from the cabinet. Its committee system permits a division of labor. The committees, composed of popularly elected ministers and top-level civil servants, bring together ministries involved in the same policy sphere. Superministries are a third innovation for the postwar British executive. The Labour government created the Department of Science and Technology in 1965 to bring under one "overlord" the Atomic Energy Authority as well as the Department of Scientific

45. See Anthony King, *The British Prime Minister* (London: Macmillan, 1969), especially parts 2 and 3 with their interviews with Harold Wilson and American political scientist Richard Neustadt.

and Industrial Research.[46] The Conservatives created the Department of the Environment, which includes the Ministry of Housing and Local Government and the Ministry of Transport. Sparked by the energy crisis of 1973–74, the Conservatives created a new superministry, the Department of Energy.

Each innovation is intended to increase cabinet-level coordination. But they may make it more difficult to detach policy from ministerial preoccupations. For example, land-use policy has been the object of great interest for a variety of groups including realtors, contractors, national park officials, urban planners, and even the Ministry of Defense, which owns large tracts of land. Lower-level civil servants precede any decision with intricate consultation with outside groups. It may take months to reach a mutually satisfactory agreement. At that point it is very hard for ministers or superministers to reject such an agreed solution, let alone devise an innovative solution of their own.[47]

One force coordinating environmental and other policies is the vision of the current prime minister—whether it is Heath's nonelite toryism with its emphasis on economic efficiency and competitiveness or Wilson's pragmatic socialism with its drive for redistributed wealth but an eye on the opinion polls. Another coordinating force is the bureaucratic power of the Treasury. Yet despite these factors, one former cabinet minister can refer to the British government's policy process as "the Whitehall obstacle race." [48]

46. For a detailed analysis of the reorganizations that accompanied post-war changes in British science-government relations, see Norman J. Vig, *Science and Technology* (London: Pergamon Press, 1968) and Vig, "Policies for Science and Technology in Great Britain: Postwar Development and Reassessment," in Christopher Wright and T. Dixon Long, eds., *Science Policies in Industrial Nations* (New York: Praeger, 1975).

47. Richard Rose, "Politics in England," in *Comparative Politics Today*, ed. Gabriel Almond (Boston: Little, Brown, 1974), p. 156.

48. Ibid., p. 177.

### NUCLEAR-REACTOR CASE

One such Whitehall "obstacle race" has concerned nuclear-reactor policy. At the outset the decision appeared to be strictly technical, remaining within the confines of ministries, as do most British policy decisions. But by 1974 nuclear-reactor questions had burst out into the public arena and seemed to be heading for resolution at the cabinet level.

Early in 1974 the House of Commons Select Committee on Science and Technology released a report based on months of hearings recommending that the government not purchase American-developed nuclear reactors. Instead, the committee urged that the government take more time and use British-designed gas-cooled reactors. The rationales were economic as well as environmental. In light of the nation's serious trade deficit British, not foreign, reactors should be the basis of any nuclear energy program. Furthermore, the Members of Parliament expressed skepticism toward the safety of the American lightwater reactor designs, despite the U.S. Atomic Energy Commission's assurances.

Though the committee did not have final authority in the matter, its report did halt the political momentum of the Central Electricity Generating Board (CEGB) which produces electricity for the nationalized system of England and Wales. The board was pushing the secretary of state for trade and industry (himself an M.P.) and the new secretary of state for energy to make a decision in favor of importing the American reactors. The CEGB had powerful interest groups on its side, including the National Nuclear Corporation, a private-public body dominated by the General Electric Corporation. The CEGB and the GEC shared the advantages of official access and status as well as the public's anxiety over the energy shortage.

On the other side of the reactor debate were important groups with policy influence stemming from positions in the policy process. The Atomic Energy Authority had invested twenty-five years of work in development of the gas-cooled reactors. Initially the AEA naturally resisted the CEGB proposal to purchase American designs. But later AEA director Sir John Hill backed down at least to the point of not vetoing the American purchase if it were combined with a government commitment to continue the research effort of his own agency. More outspoken in his opposition to importing the American reactors was the government's chief scientific adviser, who had a voice in the ultimate evaluation to be made by the nuclear safety inspectorate. He was supported in his doubts by the prestigious Institute of Professional Civil Servants, representing eight thousand nuclear scientists and engineers.

Outside interests were not crucial in this policy process. Though Friends of the Earth testified before the Commons Select Committee opposing the American reactor and though labor unions expressed their hope that "buy British" would be the government's guideline, they were not the featured players in the debate. Local communities, some of which already were sites for nuclear-powered generators (with eighteen more to become reactor sites by 1983, according to CEGB projections), apparently played virtually no part in the central government's decision.

The process was more open than in many instances due to the hearings and the committee's publicized report. The final choice, nevertheless, rested in the ministries and perhaps in the cabinet if economic repercussions pushed it up to that level. The groups most capable of pressing their points of view on the Department of

Science and Technology were quasi-public boards, business firms with official access, nationalized industries, and senior civil servants.

Five months after the Commons report the Labour government, via the secretary of state for energy, announced that it would not purchase the American design but instead would rely on British technology. Implicit in the decision was a doubt about the safety of the American AEC-backed lightwater reactors. More explicitly, there was the government's accompanying announcement that in the future the British General Electric Company, an ardent supporter of the American design, would have a less influential role to play in such energy decisions.[49]

## BUREAUCRATIC POLITICS

Policy initiatives in Britain come from the ministers appointed by the prime minister and from their departments, which, more than American executive departments, are run by career civil servants. British civil servants have long enjoyed a public prestige that would make American bureaucrats envious. But the British civil service has been the object of unusual criticism lately because of the apparent inability of government to cope with the problem of a sagging economy. Debate about the "bureaucratization" of British public life was prompted by the release of the 1968 Fulton Committee Report, which questioned the reliance on bureaucrats for policy

49. This account is based on Brenda Maddox, "Nuclear Power: Britain's Choice," *Washington Post*, 10 March 1974; "So What About Nuclear Power?" *Economist*, 9 February 1974, pp. 62–63; "The Nuclear Reactor Row," *Economist*, 26 January 1974, pp. 74–79; *New York Times*, 5 February 1974; Select Committee on Science and Technology, *The Choice of a Reactor System* (London: Her Majesty's Stationery Office, 1974); *New York Times*, 11 June 1974; "Nuclear Power: Buying American," *Economist*, 27 October 1973, p. 96.

making and made proposals to restore party responsibility. For "party government" is difficult to practice when the political fact of life is that whenever "a new party comes to power in Britain, only about a hundred politicians move into Whitehall to run the 800,000 civil servants." [50]

The most influential civil servants are the 1,169 who hold the rank of assistant secretary or above. Their power is enhanced by the practice of senior ministers delegating relatively little responsibility to their parliamentary and party colleagues serving as junior ministers.[51] This may explain why junior ministers find themselves an unequal match for the chairmen of the nationalized industries they allegedly supervise. Despite their influence over policy creation and implementation, civil servants are protected by a tradition of anonymity and confidentiality which reduces party control over the machinery of government and obstructs extragovernmental groups trying to monitor environmental policy. Politicians rarely intervene in official recruitment or promotion: "The officialdom administers itself." [52] To preserve their anonymity career officials' role in shaping a given policy is never revealed in Parliament or the press. This leaves their ministers— "their masters"—to take both credit and blame for government decision, although decisions may have been hammered out by the middle-level bureaucrat in his department or in an interdepartmental committee. When the insulating effects of bureaucratic hierarchy and continuing recruitment of civil servants from upper-class schools are added, it is clear that Britain's policy process is anything but open.

50. Anthony Sampson, quoted in Michael R. Gordon, "Civil Servants, Politicians and Parties: Shortcomings in the British Policy Process," *Comparative Politics* 4, no. 1 (October 1971): 47.

51. Gordon, "Civil Servants," pp. 44–47.

52. Richard E. Neustadt, "White House and Whitehall," in King, *British Prime Minister*, p. 135.

Although the British civil servant is more neutral than his Japanese counterpart, Labour party ministers have found the entrenched power of the central bureaucracy especially frustrating. Former Labour minister Barbara Castle recalled:

> I thought, looking back at the Labour government, how effectively the civil servants impeded us by saying we could not do some of the things our successors are now doing with remarkable facility.[53]

Another Labour minister reported a similar sense of conspiratorial opposition:

> I knew there was this inner committee of permanent secretaries. The basic economic strategy of the government was being planned by this collection of civil servants. How rarely could we ever have a discussion in the cabinet without it virtually having been made a *fait accompli* by the previous decisions behind the scenes.[54]

The bias in bureaucratic recruitment continues to favor liberal arts graduates in an era of increasingly technical demands on policy makers. Civil servants who entered government between 1948 and 1960 and now command senior posts are characterized by a remarkably low proportion—a mere 4 percent—of entrants with mathematics or science degrees. According to critic Clive Irving, "This means that 96 percent of the Whitehall leadership has the classic arts background of the civilized amateur—Greeks wrestling with a Roman world." [55]

---

53. Clive Irving, "Whitehall: The Other Opposition," *New Statesman*, 22 March 1974, p. 383.

54. Ibid. Perhaps the most revealing description of a British cabinet minister's efforts to come to grips with his civil servants is included in Richard Crossman's *The Diaries of a Cabinet Minister* (London: Jonathan Cape, 1975).

55. Irving, "Whitehall: The Other Opposition," p. 383.

Among those "Roman world" problems that this bureaucratic generation has had to cope with have been such technical complexities as the Concorde, the third airport for London, the Channel tunnel, and North Sea oil drilling. In each instance, the problem outlasted party governments while the longer-tenured civil servants remained. The Concorde already has outlived three Tory and three Labour governments and six ministries. While politicians come and go, businessmen with immediate stakes in these projects learn to develop stronger ties with the durable civil servants in either the Department of Trade and Industry or the Department of the Environment. In order to facilitate the relationship, the British businessmen tend to adopt the civil servants' style of policy making, sloughing off the normal business ethic of rugged individualism.[56]

Yet the infusion of more technically trained specialists may provide little satisfaction for environmentalists calling for a more open political process in Britain. A survey of British civil servants, for instance, found that the specialists now in government service were "less tolerant of political influences on policy, less whole-heartedly egalitarian and libertarian, less comfortable in the gray areas between politics and administration."[57]

Interdepartmental conflict exists among British bureaucrats, but there is a much stronger incentive in Britain than in the United States for bureaucrats to keep disagreements below cabinet level and thus reduce the opportunities for political leaders to intrude on bureaucratic processes.[58] Whereas the American bureaucratic process is marked by personal and institutional insecuri-

---

56. Ibid., p. 384.
57. Robert D. Putnam, "The Political Attitudes of Senior Civil Servants in Western Europe: A Preliminary Report," *British Journal of Political Science* 3, pt. 3 (July 1973): 285.
58. Richard E. Neustadt interviewed by Henry Brandon, "10 Downing Street," in King, *British Prime Minister*, p. 128.

ties which make bureaucrats, political appointees, and Presidents uncertain about their authority and the fate of their favored programs, a tacit treaty between British politicians and civil servants ensures that ministers will consult their careerists and respect their professional autonomy in exchange for the assurance that civil servants will conscientiously carry out ministerial and cabinet instructions without regard to personal partisan or ideological opinion.[59] The effectiveness of the environmental protection system that has been devised in Britain depends on the ability of civil servants to utilize discretion within policy limits set down broadly by their political "masters." [60]

Interdepartmental competition between various parts of the bureaucracy occurs over funds, prestige and programmatic priorities. Weaknesses at the top of the British executive as well as the overlapping of policy areas exacerbate this normal competitiveness.[61] Three mechanisms instituted to counter interdepartmental conflicts are: (1) the expansion of the staff resources available to both the prime minister and the cabinet secretariat, (2) a broader definition of the coordinating authority of the Treasury, and (3) the creation of superministries such as the Department of the Environment.

Civil servants play a more important role than do ministers in coordinating the operations of the central government. Ministers are likely to remain silent at cabinet meetings on matters outside their field of responsibility. One former minister has said "The one thing that is hardly ever discussed [in cabinet] is general policy." [62]

59. Neustadt interviewed by Brandon, "10 Downing Street," pp. 128–29.

60. For a discussion of this discretionary field stemming from common law, see D. A. Bingham, *The Law and Administration Relating to Protection of the Environment* (London: OYEZ, 1973).

61. Paul K. Mackal, "Trends in British Bureaucratization," *The British Journal of Sociology* 23, no. 1 (March 1972): 67.

62. Rose, in Almond, *Comparative Politics Today*, p. 179.

Bureaucratic committees that shadow each of the cabinet policy committees facilitate interdepartmental coordination. The Treasury also acts as a primary guarantor of executive coordination. Since the war the Treasury has evolved into an institution for overall steering of the economy. Its basic authority is "formidable," since every major department expenditure must obtain prior approval from the Treasury.[63] On the whole, however, Treasury's power is negative and its methods are consensual. Forceful ministers, such as Peter Walker, the first secretary of state for the environment, can override the Treasury officials by taking an expenditure question straight to the cabinet for resolution. Yet "overlords" of the Department of the Environment have a special interest in more assertive coordinating efforts, since environmental issues are by nature so diffuse. Peter Walker, in fact, warned during one parliamentary debate that future chancellors of the exchequer would have to give more attention in budgetary planning to the importance of natural resources planning and to such fundamental questions as the objectives of growth versus environmental preservation.[64]

63. Beer et al., *Patterns of Government* (3rd ed.; New York: Random House, 1973), now-1st mention, p. 188. Richard Neustadt takes exception to the perhaps wishful assumption of American budget officials (in Bureau of Budget or its successor Office of Management and Budget) that they are genuine counterparts of the British Treasury officials. Neustadt points out that American "budgeteers," unlike the British, are *non*careerists. They hold jobs "infused with Presidential interest or concern"; they are either "in and outers" who come from and return to law firms, banks, universities, or "up and outers" who give up their civil service status to move on to cabinet or other political appointments. They are the administrative elite who share in governing. By contrast, the British Treasury official is among the intellectual and moral elite of Britain; he can afford to be dispassionate in his arguments and lose with grace. He is "a disciplined man, but a man fulfilled, not frustrated. His discipline is what he pays for power." Neustadt, "White House and Whitehall," pp. 135–36.

64. *Times* (London), 24 March 1972. In their study of Britain's budgetary politics, Hugh Heclo and Aaron Wildavsky note that the budget is still the result of extensive bargaining between Treasury

## DEPARTMENT OF THE ENVIRONMENT

Prior to the launching of the Department of the Environment in 1970, control authority for environmentally related policy was scattered throughout a dozen government departments. Air quality alone was handled by the ministries of Housing and Local Government, of Technology, of Transport, of Agriculture, the Department of Social Services (health), the Board of Trade (factories), and the secretaries of state for Scotland and Wales. The burden on interdepartmental committees was awesome. They were aided somewhat by the Clean Air Council, under the Ministry of Housing and Local Government, and the Clean Air Council for Scotland, subject to the Scottish Development Department whose members were drawn from inside and outside government. Industrial pollution was supervised by the Alkali Inspectorate reporting to the Ministry of Housing and Local Government and to the secretaries of state for Scotland and Wales.

Conservative and Labour parties claim credit for establishment of the first cabinet-level executive department dealing explicitly with environmental affairs. Neither Japan nor the United States has such a department; nor does the Soviet Union. In Britain, the public was already pollution conscious and a number of antipollution laws were on the books.

Sensing an issue that would enable him to initiate a major revision of governmental machinery which would promote two other objectives Labour backed—regional and urban planning—Prime Minister Harold Wilson proposed to Parliament in 1969 a consolidation of func-

officials and individual ministries, but that Britain has gone further than many countries in instituting effective budgetary planning. Heclo and Wildavsky, *The Private Government of Public Money* (Berkeley: University of California Press, 1974).

tions.[65] Out of this came the Ministry for Local Government and Regional Planning, responsible for investigating "environmental pollution in all its forms and to make proposals to me [Wilson] on how this should be dealt with." However, Wilson let the existing Ministries of Housing and Local Government and of Transport continue. It was one step short of a genuinely integrated superministry.

There was speculation at the time as to Wilson's true motives in selecting Anthony Crosland as the first secretary of state for Local Government and Regional Planning, since the new ministry "seemed to resemble those power-stations in Outer Siberia where erring members of the Soviet Praesidium are from time to time despatched." One clue as to the dubious political status of the new post—and thus of the environment as a political priority —was the difficulty one had in even contacting the new "overlord." Nora Beloff, political correspondent for the *Observer*, recalled: "I telephoned the Ministry of Housing and Local Government. The name seemed unfamiliar to them. I was given another number where the operator announced herself: 'Department of Economic Affairs'— ten days after the Prime Minister had formally abolished it. 'Crosland?' he said, 'Mr. Crosland? You must mean *Miss Edith* Crosland in research. . . .' I found him in the end in the ill-fated office once occupied by Mr. George Brown [a former Labour minister]." [66]

When the Conservatives upset the Labour party in the 1970 elections, they introduced the issue of environmental protection for the first time into the Queen's Speech at the opening of Parliament and pressed forward the administrative consolidation begun by Wilson. But

---

65. The following material concerning the evolution of the Department of the Environment is drawn from Stanley P. Johnson, *The Politics of the Environment: The British Experience* (London: Tom Stacey, Ltd., 1973), pp. 94–130.

66. Ibid., p. 98.

there were still skeptics who doubted that any such department could actually be formed. In October 1970, Prime Minister Heath presented to Parliament a White Paper on "the Reorganization of the Central Government" in which Heath made it clear that among his principal priorities was to bring more tough-minded efficiency into British governmental operations. The Department of the Environment (DOE) would be responsible for regional infrastructure and services, though the Department of Trade and Industry would lead the way in regional economic development. Secretary of state for the environment would have under him the minister for local government and development, the minister for housing and construction, and the minister for transport. None of these three ministers would serve on the cabinet, but would be represented there by the new secretary of state.

When pressed in Commons to explain how the new secretary of state for the environment could possibly hope adequately to supervise such a vast array of functions, a spokesman for the government acknowledged that detailed control would be impossible. But, he explained, the creation of the secretary of state and his staff amounted to the creation of an in-house "lobby" which would, for instance, act as a counterweight to a Department of Trade and Industry proposal which supported airport expansion but without giving attention to noise pollution. The secretary of state for the environment would act as a lobby "in favour of the care of people's rights.[67] The actual capacity of any appointee to carry out this mandate remained in doubt since, as one journalist commented, "Whitehall was like the Balkans—nobody would ever draw an entirely satisfactory map."[68]

---

67. Mr. David Howell, parliamentary secretary for the Civil Service Department, quoted in Johnson, *The Politics of the Environment*, p. 130.
68. Ibid.

The fact that the Department of the Environment over the next three years did manage to hold its own both in the bureaucracy and in the cabinet is due to: (1) its public visibility as a result of newly vigorous environmental groups with a stake in DOE's decisions; (2) Prime Minister Heath's selection of one of his own closest political allies for secretary of state; and (3) the personal attitudes of that secretary of state, Mr. Peter Walker, who brought energetic drive and political assertiveness to the office. On the other hand, Walker and his successors at DOE headed a department that was divided in its mandate, since it was not only the protector of the environment, but the promoter of road and housing construction which often occurred at the expense of environmental preservation.

At first glance Peter Walker hardly seemed the auspicious choice for leading environmental affairs. The British press portrayed him as the "epitome of the new dynamic thrusting Tory. A man who had started life as an office boy, who had borrowed £200 when he was 22 and built up an insurance empire worth at least £1,500,000. . . . A man who claimed he could manage on only five hours of sleep a night, which gave him at least 90 minutes start on everyone else in the City [London's "Wall Street"]. . . . National Chairman of the Young Conservatives, youngest member of the Shadow Cabinet and youngest member of the Cabinet . . . a go-getter who, at the age of 38, got there." [69]

Rather than making the DOE a puppet in the hands of British business interests, these qualities inclined Peter Walker to be impatient with British business executives who claimed that antipollution regulations would harm their companies. Instead, he urged them to manage their firms more efficiently so that British companies could

---

69. Johnson, *The Politics of the Environment*, pp. 135–36.

compete in the tough international markets without depending on government protection or subsidies. DOE benefited from Walker's style complementing that of the prime minister. Heads of environmental agencies in so many other countries have been plagued by their remoteness from the political leader. Too frequently they have been chosen to appease competing or minor factions of the ruling alliance.

When Heath took over leadership of the Conservative party, he became the first Conservative prime minister to come from a nonelite background. In this sense he was like Tanaka in Japan. Heath elevated along with him four other ambitious "climbers" who shared this lower-middle-class and upper-working-class background: Anthony Barber, Margaret Thatcher, Peter Walker, and Geoffrey Rippon. Walker was not only Heath's first environmental overlord, but one of his key campaign strategists. When in 1973 Heath transferred Walker to the Department of Trade and Industry, he was replaced at the DOE by Geoffrey Rippon. Rippon had been in Heath's cabinet since 1970. Heath and his "Heathmen," as they were labeled in the press, were distinguished from an older style of Tory leader by their drive and self-confidence as well as their disinterest in formal ideology. They were impatient with the typical search for consensus and relished confrontations. It was to be this orientation that finally undermined Heath's political support during the 1974 miners' strike.[70]

Among the obstacles facing Peter Walker was a Department of Trade and Industry preoccupied with economic growth and eager to subsidize mineral explorations in the national parks. He dealt with the Ministry of Agriculture and Fisheries which traditionally thought first

---

70. For a lively portrait of Edward Heath and his closest associates, see Andrew Roth, *Heath and the Heathmen* (London: Routledge & Kegan Paul, 1972).

of food production and placating farming interests rather than of environmentally determined land-use planning or control of pesticides. Local government officials fought his moves to impose planning guidelines. Even within his own DOE, there were highway and traffic engineers who seemed to be consolidating their own positions through ties with local government engineers. And the Walker-supported Skeffington Report calling for more public participation in planning processes, not surprisingly, met with little enthusiasm from his civil service subordinates.[71]

Walker admitted fascination with American politics.[72] In that regard he was criticized for doing things for effect. But he had the respect of his civil servants, a crucial property for any British minister. They told newsmen that they were impressed with his self-confidence and quickness. However, "like a nervous crowd watching an acrobat dancing along a high wire, they rightly fear the crash and marvel at his survival." [73] Part of that survival required keeping the peace among the several ministers under him. He met every morning in his sixteenth-floor office at the DOE with his ministers. Civil servants were excluded. Relations among the ministers and between them and the secretary of state were not always harmonious.

Walker pushed his civil servants to do things they thought they could not do. He delegated more authority to his junior ministers and instructed civil servants to go to those junior ministers for important decisions, instead of

---

71. Malcolm MacEwen, "Mr. Walker's Slip Is Showing," *New Statesman* 83, no. 2144 (21 Apr. 1972): 520. For the environmentalists' view of these proposals, see "Comment: Open Government or Open Contempt?" *Ecologist* 3, no. 2 (February 1973): 47.

72. Walker became a close friend of Senator Robert Kennedy and attended every Republican and Democratic party convention between 1952 and 1972. *Times* (London), 10 May 1972.

73. Ibid.

treating them as "dogsbodies" (private secretaries). Walker also utilized patronage powers to bring into the department people in their twenties. His innovations were directed toward closing that gulf which had so long separated ministers from civil servants in the British system.[74]

In his three years at DOE, Walker's chief accomplishment was the institutionalization of an integrated department in the face of strong tendencies toward bureaucratic "balkanization." In addition, he shepherded through the policy process far-reaching local government reforms which would facilitate national planning and more effective antipollution administration.[75]

Whether Walker's accomplishments will last is open to question. His successor, Geoffrey Rippon, while also a close political associate of the prime minister, did not have Walker's flare. When Labour returned to power after the 1974 election, Harold Wilson did not select one of his party's strong men to take over DOE. He appointed Anthony Crosland as secretary of state for the environment. Crosland is an astute Labour politician with past experience holding other portfolios. Crosland reflects his party in his particular interest in land reform as the chief lever for affecting fundamental change in British social and environmental relations. However, control of land speculation is a highly controversial issue in Britain and could preoccupy the new head of DOE, a sprawling department which needs firm leadership if it is not to dissipate its energies.

---

74. Ibid. British environmental activists agreed that Walker was more forceful than Rippon, but voiced skepticism about his real impact and the depth of his environmental commitment. Some characterized Walker as merely "trendy," a lot for show but few hard decisions. Personal correspondence to the author from Julian Cummins, Secretary General of The Youth Federation for Environmental studies and conservation, 20 August 1974.

75. A description of the Walker strategy for dealing with his

On the Conservative bench, too, there was subtle downgrading of the environmental post. In Heath's "shadow cabinet" the environmental post went to a third "Heathman," Margaret Thatcher. But it was reported that she had requested that position "as a break" after three and a half stormy years as minister of education.[76]

A further change could undermine the efficacy of the DOE in bureaucratic policy processes. In the wake of the oil crisis the Heath government established a new super-ministry, the Department of Energy, headed by Heath's closest adviser, Lord Carrington, who in turn recruited "Heathmen" for the top posts. The immediate loser was the Department of Trade and Industry, which surrendered control over coal, electricity and offshore oil.[77] In the longer term, however, the greater loser might turn out to be the Department of Environment. It now faces even stronger forces urging the primacy of growth over environmental protection.

## CONCLUSION

Britain's centralized political system has allowed it to engage in long-range growth planning to an extent unknown in more fragmented systems such as the United States. The environmental issue never has mobilized as vital a movement as in either the United States or Japan; but environmental questions have met with significant governmental response despite this limited mobilization, thanks to centralization in London and to the professional prestige of Britain's civil service.

subordinate ministers in the newly formed DOE can be found in Tony Aldous, *Battle for the Environment* (Glasgow: William Collins Sons, 1972), pp. 13–31.

76. "Shadow Cabinet: Who Are Tomorrow's Men?," *Economist*, 16 March 1974, p. 21. In February 1975 Margaret Thatcher, Heath's chief rival for Conservative party leadership defeated him.

77. See *New York Times*, 9 January 1974: "The DTI Dismembered," *Economist*, 9 March 1974, p. 78.

The priorities in environmental planning have been control of urbanization, especially the preservation of green belts around major cities and the reduction of popular migrations from the north to the southern metropolitan region around London.[78] Because Britain's environmental issues have been so intimately related to the problems of urban growth, their solutions have taken the form of reorganization of local governments and a merger of housing, local government, and transportation ministries into the new superministry for the environment. This superministry permits the British government closer coordination of environmental policy making than enjoyed by most other countries, though the secretary of state for the environment still needs political stature within his own party plus an awareness of the dynamics of British bureaucratic politics to ensure that paper authority is translated into genuine policy influence.

The conventional model of British politics that has had such an impact on ex-colonies around the world does not fit the contemporary reality. Environmentalists in Britain fall into two rough categories: (1) those who remain comfortable with a system that relies on private consultation among established interest groups and power holders and shies away from abrasive confrontations or sanctions, and (2) those who have been importing American activist handbooks and are impatient with this exclusivist and often collusive sort of political process and feel that many new interest groups deserve more access into policy circles. But both would agree that the model of party government and collective responsibility scarcely fits the political reality in which environmentally relevant decisions are made by civil servants, in which

---

78. Jack Underhill, "Great Britain Revisited: Some Thoughts on New Towns, Urban Planning and Growth Policy," *HUD Challenge* 5, no. 7 (July 1974): 18–23. See also Stephen L. Elkin, *Politics and Land Use Planning* (New York: Cambridge University Press, 1974).

the prime minister is far more than the first-among-equals, and in which average citizens feel their local governments are of little importance and the major parties are less and less satisfactory.

Environmental politics itself will probably not be central enough to force Britons consciously to shape a new model, but when its dynamics are combined with those of national economic deterioration, it may play a crucial role in such a political reformation. It is entirely possible that in Britain, as in some Third World countries as well, the lack of rapid economic growth will give citizens a greater sense of what is to be sacrificed if growth is urged mindlessly.

CHAPTER 9

# CONCLUSION

We began this investigation with two questions. First, are there significant variations in the ways different political systems handle environmental problems? We were interested in the ease or difficulty that environmental matters faced in gaining status as a legitimate political issue. In addition, we were concerned with the structural and cultural factors that shaped the strategy of governments when tackling environmental hazards. Our second question related to the broader scope of politics: what do the various cases of environmental policy making tell us about political systems as a whole? In pursuing this line of inquiry we were acknowledging that particular policy areas may not be true reflections of the given nation's political system in all its dimensions, but environmental

**317**

policy processes might alert us to neglected attributes and potential changes in particular systems.

Surveying the conditions shaping the political handling of environmental problems, one is struck perhaps as much by the similarities as the differences. The differences are most apparent in the role that ordinary citizens can play in formulating public policy and in the sorts of strategies that reap the greatest policy rewards in various nations. But beyond these strategic differences that derive from institutional variations and nongovernmental organizational resources, there are important shared attributes. Perhaps the most interesting and neglected is the intimate relationship between public and private power or economic and political management in most countries, a relationship which biases most governments against taking decisive action to limit man-made environmental hazards.

Because the dichotomy between socialist and capitalist systems has been so widely used in comparative research, one is not prepared for the constant blurring of lines one finds here between "government" and "business" in so many contemporary systems. In socialist systems factory managers and their ministerial supervisors are as concerned about high productivity as their counterparts in the private sectors of capitalist countries. Personal *careers*, professional *training*, and technical *measures* of organizational efficiency all make them reluctant to take account of the ecological consequences of unrestrained economic expansion. Then, too, in noncommunist systems there is a growing tendency to involve government in the regulation or subsidization of commercial enterprises. Electrical, petroleum and nuclear power utilities, airlines, railroads, highway construction, dam projects, and military weaponry are typically under direct government authority.

Indirectly, as well, governments bolster certain enterprises by granting public contracts, loans, and licenses. Likewise, there is the increasing reliance of governments on "mixed" or quasi-public boards for developing environmentally related economic programs. In the U.S. alone there are now 18,000 government-created authorities monopolizing certain commercial spheres and giving priority to their profit margins and not to the more abstract notions of public welfare.[1] Usually membership on such boards is restricted to government administrators and business-affiliated experts. Even when the quasi-public board is depended upon to regulate a business sector, membership may be drawn chiefly from firms with an interest in diluting regulatory restrictions.

Finally, because of the limited pool of expertise in most political systems there is a common rotation of technically trained personnel between private and public offices, frequently without a parallel alternation of role perceptions so that the distinction between public and private well-being not only is blurred structurally, it is blurred in the minds of individual decision makers themselves. To describe and explain the dynamics of environmental politics in a variety of nations accurately, therefore, we need to go beyond gross distinctions and look at the real operational relationships between political and economic power in each case.

It follows that the governments we have discussed rarely stand as neutral umpires in environmental controversies. The image of a national government as an arbitrator buffeted to and fro by opposing nongovernmental forces disguises the important self-interests that governments possess. We have seen how most modern governments and modernizing governments as well have

---

1. Two perceptive studies of the factors that insulate such authorities against pressures of environmentalists are Dorothy Nelkin, "Massport vs. Community," *Society* 11, no. 4 (May/June 1974): 27–39; James W. Hughes, "Realtors, Bankers and Politicians in the New York/New Jersey Port Authority," *Society* 11, no. 4 (May/June 1974): 63–73.

strong stakes in economic growth. One reason that certain bureaucratic agencies wield more power than others in a given country is that they have a closer connection to that governmental self-interest. Should low levels of bronchitis come to replace high levels of GNP as a measure of a government's power, then perhaps health ministries and environmental agencies will find that they are on firmer ground in intragovernmental debates. It is important, consequently, to look not only at conflicts between parties or interest groups in a country, but also at conflicts *within* government over the definition of governmental self-interest.

Another area of perhaps unexpected similarity has been between developed and underdeveloped countries. While one should certainly not underestimate those limitations and dilemmas facing Third World nations that reduce the saliency of pollution questions, it is simplistic to assume that development progresses in such an orderly fashion that pollution will be saved for highly industrialized countries. In reality, countries in Africa, Asia, and Latin America, which still fall well below protein requirements and are only just instituting compulsory primary schooling, are simultaneously confronting the "postmodern" challenges of urban sprawl, supertanker leaks, deforestation, and traffic-produced smog.

Similarly, within allegedly "developed" nations there are sectors of notable underdevelopment. If one defines development in terms of *the capacity of the system to cope with the demands upon it,* then it is clear that countries as affluent as Germany and Japan and the United States fall short of genuine development. Furthermore, even using the more conventional definition of development, a great many industrialized nations contain social and regional groups that have not shared in the fruits of economic affluence. These groups and the politicians depending on their support are especially wary of environmental controls which threaten to halt growth just at the time when they are about to share in its rewards. Furthermore, growth-oriented national regimes look to underdeveloped territories within their boundaries (Scotland, Alaska, Siberia, Hokkaido) for new sources of energy and new outlets for population growth.

Comparative analysis also suggests that, regardless of structural differences, all modern political systems give prominence to those affairs that can most easily be measured. What can be *measured* can be converted into an issue, a goal, a criterion for performance. What cannot be measured readily or precisely will have an uphill battle in achieving political centrality.

Environmental hazards and long-range environmental well-being have been most politicized (1) where systems are equipped with technical experts who can translate conditions of nature into comprehendable societal dangers, (2) where those experts enjoy political access, (3) where economists are able and willing to convert physical data into economic costs and benefits, and (4) where literacy and popular media exist so that ordinary citizens can mobilize around these newly defined dangers. Ideology, culture, levels of modernization, and incidence of dramatic catastrophes all effect the distribution of these conditions. Sweden and Japan far outdistanced Brazil and the Soviet Union in their politicization of environmental hazards because they are more willing and able to employ *measures* which highlight the public implications of those hazards. One reason why understanding of environmental politics' varying patterns requires a sensitivity to budgetary processes is that in hammering out a government budget the biases in favor of certain measures and dismissal of others become clearest.

The major differences between countries in their political approach to environmental decision making coexist with these significant similarities. We have seen that political parties are critical in some environmental political processes while peripheral in others. Parties are most important when a single party has dominated the political system during a long period of economic growth. In Japan, the USSR, Sweden, and Canada's province of British Columbia the party in power has become so intimately wedded to economic definitions of public good that doubts about that definition also raises doubts about the relevance of the dominant party. The full impact of environmental issues will be felt on the party system, however, only where, as in Sweden and Japan, effective opposition parties are capable of exploiting the environmental discontent in elections or in grass-roots organizing.

This suggests a second important dimension along which political systems have differed in their environmental policy processes: local government. Because the harmful effects of pollution are usually felt in specific localities, those political systems in which local governments possess the greatest authority and elicit most citizen interest are likely to experience the greatest politicization of environmental issues. On the other hand, where local governments are weak or have been allowed to atrophy, environmental movements, once mobilized, are apt to give new vitality to community politics. Only recently have political scientists paid attention to local politics outside the United States. If the differences between countries' handling of environmental problems is to be fully understood, we will have to devote far more attention to comparative *local* political analysis.

A third source of variation has been the unequal spread among nations of the resources for citizen mobilization, especially mobilization autonomous from regime control, and for citizen policy access. Dozens of comparative studies have shown that literacy, affluence, urbanization, cultural legitimation—all affect levels of citizen mobilization. In addition, the analysis of environmental politics demonstrates that citizen access to scientific expertise and governmental fragmentation providing a multitude of alternative access points both enhance the launching of an effective citizen movement.

Concerning our second line of inquiry—what environmental politics can tell us of the larger political system—it is necessary to spell out the distinctive qualities of the environment as a policy matter. These distinctions will narrow the generalization we can extract from our survey, but they should avert some common pitfalls in comparative analysis.

The foregoing chapters suggest the following distinctive characteristics adhering to the environment as a political policy area:

1. The boundaries of the issue are constantly expanding as technical knowledge becomes more refined; it is more meaningful today, for instance, to refer to the "politics of the environment" rather than merely the "politics of pollution."
2. The constituencies for environmental questions are unusually motley in that they include potentially all human beings.
3. Environmental controversies have a way of touching fundamental cultural and ideological nerves in most systems.
4. As an issue, environmental disruption exhibits considerable fickleness, being constantly challenged by policy matters that are more subject to concrete measurement and are more related to citizens' immediate anxieties over jobs, cost of living, or foreign enemies.
5. Environmental issues are especially hard to confine neatly within nation-states borders; a

domestic pollution question quickly slops over into the international arena.

6. The time span necessary for effective policy formation is more drawn out than in other policy areas.

7. Resolution of environmental problems puts especially heavy burdens on government in terms of scientific expertise and technical monitoring facilities.

What makes analysis of environmental politics so fruitful to the generalist is the likelihood that these very distinctions will become characteristic of many other fields of public policy in the future.

One cannot help but remark on the international impact of one lone political figure: Ralph Nader. Political activists in countries as diverse as Britain, Japan, and Australia either have invited him to come on inspection tours or deliberately borrowed his organizational techniques or labeled certain of their own spokesmen "Ralph Naders." In the United States, Nader and his various spinoff units have had an impact because they have focused an analytical spotlight on those corners of the American political process usually neglected by domestic observers: the corporation and the bureaucratic agency. Outside the United States, Nader's influence has been greatest with regard to modes of citizen mobilization. But in both instances it is the special characteristics of environmental politics that have given the "Nader formula" such appeal, especially its demand for information as a political resource. The Nader formula, while nowhere spelled out concisely, consists of the following:

1. a commitment to openness in policy processes, openness in terms of distribution of pertinent *information* and in terms of citizen *access*

2. faith in an *adversary* mode of decision making, a wariness about any organization being granted sole authority to speak for the "public interest"
3. a complementary belief in *competitive economics* and a debunking of the assertion that concentration ensures efficiency in any market sector
4. perception of the citizen as *rational*, able to make connections between self-interest and public interest if supplied with adequate and accurate information
5. reliance on *legal sanctions* and the courts to ensure corporate and bureaucratic accountability

There is a paradox in the Nader formula. It is addressed to the American polity's shortcomings; yet it is profoundly American in its style and cultural assumptions, especially its faith in a rational citizenry, the potency of information, and the preference for adversary relationships. The attraction that the Nader organizational techniques has had in other countries suggests that some of these heretofore distinctively American political attitudes may gain relevance in political systems having to come to grips with the new problems of environmental decay. On the other hand, the structural and cultural differences that we have seen separating political systems will mean that the Nader formula either will be adopted only in part or in some modified fashion in other nations. What then becomes interesting is *what* variations on the Nader formula will develop between, say, Japan and Britain.

For the political generalist environmental politics furthermore underscores the extent to which political systems are underdeveloped because of (1) an incapacity to coordinate increasingly fragmented government operations and (2) an inability to engage in long-term comprehensive planning. The current international discussions

concerning inflation and monetary reform, population
control, urbanization, and food supplies all indicate that
the need for governmental planning and coordinating
capacities is not confined merely to the environmental
policy sphere. Socialist systems long have claimed planning and coordinating as their strong suits. This has been
particularly true when administrative centralization has
been combined with single party domination. But our
brief case studies suggest that even in socialist systems,
narrow organizational self-interests can take root and
eventually dilute the effects of central policy guidelines.
Moreover, for planning capacities to have their intended
ameliorating effects, the *priorities* around which plans are
designed must be accurately pegged to real national
needs. If priorities continue to be distorted by doctrinal
commitments, planning effectiveness actually may be
even more harmful in its consequences than bumbling ad
hocism.

Among the countries here surveyed the United States
is perhaps the most severely underdeveloped in terms of
planning and coordinating capacity. While Western European nations such as France, Britain, and Sweden have
been pushing ahead with regional development plans and
urban planning on a nationwide scale, the American
political system cannot digest even a very modest land-
use planning bill. The cultural preference for laissez faire
relations (and the belief that they still persist when in fact
there is increasing, though selective, government-business interlocking) and eschewance of ideological modes
of debate enshrine pragmatism and ad hoc bargaining;
while structural fragmentation in the form of federalism,
local home rule, and "subgovernment" specialized policy
clusters add further barriers to nationwide planning.

It has been environmental politics, however, which
has been the principal factor disclosing the inadequacies
of these traditional political modes. Although centralized

planning and long-range national priorities remain questionable objectives among Americans, the Environmental Protection Agency, despite its political weakness, is imposing planning guidelines just by carrying out its own mandates. To fulfill the requirements of the Clean Air Act by 1975, the EPA is requiring drastic changes in the economic and social life of the country's major cities. To meet federal water standards monitored by the EPA, towns across the country are imposing moritoriums on housing, thereby affecting national population migration patterns. If planning cannot come in by the front door, it may enter through a side door.[2]

These two consequences of environmental politics— the increasing demands for citizen access and for nationwide planning coordination—when taken together pose a developmental dilemma. For, taken in its most simplistic form, such planning and coordination imperatives usually curtail Nader-like citizen participation. Does the possible primacy of the environmental issue herald an era of increased centralization (or more firmly entrenched centralization where it already exists) and closed bureaucratized technocracy? Can long-range planning and effective harmonization of agencies coexist with political openness, decentralization, and adaptive innovation? We will need to know more about local (town, county, province) politics in most countries and about the relationships between local administrations and central decision making in order to unravel this dilemma.

A third area of political analysis in which environmental politics raises provocative questions about potential development trends is international affairs. Does environmental politics serve as a forerunner of an era when there will be little distinction between national and international systems? Our study offers a mixed answer.

---

2. See the informative three-part series by Gladwin Hill on U.S. urban growth planning, *New York Times*, 28, 29, 30 July 1974.

Take the UN-sponsored Stockholm Conference of 1972. On the one hand it was a historic testimony to the "one-world" sensitivity that has grown over the past decade. As a reflection of that awareness delegates from diverse political systems created a new UN Environmental Program with a substantial budget of $20 million for its first five years. In its second meeting held in Nairobi, its headquarters, in May 1974, the council of the nascent UNEP exhibited "remarkable harmony and a sense of common purpose," with only Brazil of the fifty-eight nations represented standing out as obstructionist.[3] Similarly, there has been a proliferation of international environment-related conferences, treaties and informal exchange relationships between nongovernmental as well as governmental organizations.

On the other hand, a great deal of the international political negotiating and cooperation is being shaped by older nation-state cultural forces. The most notable absentee at the Stockholm Conference was the Soviet Union, which stayed away in protest over the exclusion of its cold war ally East Germany. Likewise, the United States, otherwise one of the most energetic backers of the Stockholm meeting and UNEP, objected when Sweden, along with socialist and certain Third World countries insisted that biological warfare, defoliation in Vietnam, and nuclear weapons testing all be considered eminently germane to any discussion on man-made environmental hazards and their prevention.[4] In turn, several of the

3. "World Environmental Newsletter," *Saturday Review/World*, 18 May 1974, p. 35.
4. For analyses of the environmental damage caused by U.S. tactics in Vietnam, see Michael McClintock et al., *Air, Water, Earth, Fire: The Impact of the Military on World Environmental Order* (San Francisco: Sierra Clubs, 1970); John Lewallen, *Ecology of Devastation* (Baltimore: Penguin Books, 1971); Thomas Brindley, "A Legacy of Poison," *Far Eastern Economic Review*, 5 March 1973, pp. 22–23; Thomas Whiteside, *Defoliation* (New York: Ballantine Books, 1970); Peter Caplan, "Weather Modification and War," *Bulletin of Concerned Asian Scholars* 6, no. 1 (January–March 1974): 28–31.

leading Third World delegations at Stockholm subsequently charged that all this political frenzy over environmental protection was merely imperialism in a new guise, intended to preserve not the environment but the wide gap between rich and poor nations. Within the United Nations, too, the political dynamics were ones familiar to any student of traditional bureaucratic conflict. The fledgling UNEP, just like environmental agencies in Japan, the United States, and USSR, had to expend considerable energy just to fend off encroachment from better-established agencies, in this case WHO, FAO, and UNICEF, each of which beefed up its environmental programs in order to prevent UNEP from capturing more budgetary and programmatic ground.

In relations between nation-states, national interests continued to offset any new one-world consciousness:

—In 1974, France refused to stop dumping salts from its potassium mines into the Rhine River and managed to frustrate clean-up efforts by five other nations sharing the Rhine.[5]

—The International Conference on the Law of the Sea met in Caracas, Venezuela, in 1974 and found American East Coast fishermen pressing for a 200-mile territorial limit in order to stop what they claimed to be disastrous overfishing by Japanese, Russian, and Spanish fleets off U.S. shores, while American West Coast fishermen pressed for the 12-mile limit so that they could continue to fish off Mexican and Peruvian coasts.

—Brazil and Argentina engaged in a dispute over the damming of the Rio de la Plata.

—The Mexican government has protested against the

5. "World Environmental Newsletter," *Saturday Review/World*, 9 March 1974, p. 31.

U.S. government-backed damming of the Colorado River, which has resulted in such high salinity in the river when it reaches the border that farming in northern Mexico is endangered.

—Denmark's concern over air pollution focuses less on domestic sources than on the smog that travels from industrial centers of Germany.[6]

—Australians objected to Chinese and French atmospheric nuclear bomb tests in Asia.

—Canadian provincial regimes have threatened to take steps to prevent any oil refineries off the U.S. northeastern coast.

—The steady pollution of the waters off the Cape of Good Hope by Swedish, Japanese, Greek, and American-owned oil-leaking supertankers has aroused official South African protests.[7]

These disputes and dozens more of a similar character might lead one to hypothesize that environmental politics, rather than increasing international cooperation and softening national self-interests, could in fact engender more conflict and greater nationalist myopia.

The direction that international relationships do eventually take will depend in large part on two things: (1) the relative domestic influence of those *national* groups and bureaucratic agencies that are most inclined to view problems from a global, nonparochial perspective; and (2) the adaptive capabilities of existing international bodies newly confronted with environmental conflicts. In fact, environmental demands might provide existing bodies with new relevance and vitality.

NATO, for instance, has established a Committee on

6. Correspondence with the author by Joanne S. Wyman, 4 December 1973.

7. Noel Mosterat, "Profile: Super Tankers," *New Yorker,* pt. 1, 13 May 1973; pt. 2, 20 May 1974, pp. 90–94. These articles are published in a larger volume: *Supership* (New York: Alfred Knopf, 1974).

the Challenges of Modern Society (CCMS) with a mandate to work out environmental programs for member countries.[8] The Commonwealth and European Common Market and the Caribbean trade association CARIFTA also have broadened their scopes so as to take a role in environmental conflicts. These as well as a host of citizen groups which have become internationalized by environmental politics, some 250 nongovernmental groups (NGOs), took part in the Stockholm Conference and since have demanded a greater role in UN decision making.[9] Automobile and petroleum manufacturers in several industrial countries are joining forces in order to negate the trade impact of conflicting air pollution laws in different nations (e.g., the 1975 lead in gasoline requirement in France, Italy, and Japan will be "none"; in the U.S. it will be 1.7 grams per gallon and in Britain 1.8 grams per gallon by the end of 1975).[10]

Each of these real and potential trends in international and national politics demonstrates that environmental politics is more than simply a passive product of larger socioeconomic conditions. Politics itself can be a significant motor for change. Because they so frequently challenge well-established cultural norms and regime perceptions of national success and because they provoke new styles of political action and new modes of governmental operation, environmental issues are likely to continue to be an active causal factor in the political evolution of a wide variety of political systems.

8. See Charles F. Doran, "The Environmental Impact of CCMS (NATO) and the International Environmental Agency (UN)" (Paper delivered at the International Studies Association meeting, Saint Louis, March 1974).

9. Re role of NGOs, see Anne Thompson Feraru, "Transnational Political Interests and the Global Environment," *International Organization* 28, no. 1 (Winter 1974): 31–60.

10. "World Environment Newsletter," *World*, 30 January 1973, p. 70.

# INDEX

**332**